Willian Ribeiro da Silva

Between tourism and leisure. The case of the city of Panorama/SP

Willian Ribeiro da Silva

Between tourism and leisure. The case of the city of Panorama/SP

Tourism and leisure in the process of touristification

ScienciaScripts

Imprint
Any brand names and product names mentioned in this book are subject to trademark, brand or patent protection and are trademarks or registered trademarks of their respective holders. The use of brand names, product names, common names, trade names, product descriptions etc. even without a particular marking in this work is in no way to be construed to mean that such names may be regarded as unrestricted in respect of trademark and brand protection legislation and could thus be used by anyone.

Cover image: www.ingimage.com

This book is a translation from the original published under ISBN 978-3-330-75694-6.

Publisher:
Sciencia Scripts
is a trademark of
Dodo Books Indian Ocean Ltd. and OmniScriptum S.R.L publishing group

120 High Road, East Finchley, London, N2 9ED, United Kingdom
Str. Armeneasca 28/1, office 1, Chisinau MD-2012, Republic of Moldova, Europe
Printed at: see last page
ISBN: 978-620-3-81716-4

Copyright © Willian Ribeiro da Silva
Copyright © 2024 Dodo Books Indian Ocean Ltd. and OmniScriptum S.R.L publishing group

INDICE

I dedicate it to my parents, Dalton Ribeiro da Silva and Sonia Maria Mendes.

SUMMARY

Nowadays, tourism has become one of the main drivers of the world economy and is booming. This activity is characterized by the appropriation and consumption of spaces, making them valued commodities in the contemporary world, creating new territorialities. The aim of this work is to analyze the new territorialities presented in the municipality of Panorama, located in the state of São Paulo, based on the impacts caused by the formation of the artificial lake on the Parana River, with an emphasis on tourism and leisure. Panorama has undergone major transformations prompted by the construction of the Sergio Motta Hydroelectric Power Station, located in Porto Primavera, a district of the municipality of Rosana in the state of São Paulo, which complies with global logics, while its manifestations and impacts occur locally. Among the works analyzed in this paper is the Mario Covas Bridge, which connects the states of São Paulo and Mato Grosso do Sul and has created new flows. These works were analyzed according to Milton Santos' theory of engineering systems. Several factors indicate that Panorama has great potential for tourism.

Tourism and leisure have become one of Panorama's main economic activities. This activity, idealized by the managers of the territory and by the Energy Company of the State of São Paulo (CESP), has been boosted by *marketing*, the creation of tourist areas and their sale. This activity has shown itself to be perverse and contradictory, both economically and socially. It could be the way out of development, but it could also be the door to increasing inequality. In order to prepare this work, we carried out fieldwork, bibliographical research, and a temporal and spatial analysis through the municipality's history and photographs. Through this theoretical and methodological approach, it is possible to understand the new territorial dynamics that have taken hold in this municipality and to analyze the mitigating policies implemented by CESP, their relationship with tourism and leisure and the territorial "marks".

Keywords: Tourism, Engineering Systems, Territory, Panorama- SP, UHE Engenheiro Sergio Motta.

1 INTRODUCTION

In Brazil and around the world, tourism has become one of the main economic activities, boosting the tertiary sector, appropriating spaces and transforming them into commodities. The main point of this paper is to understand the new territorialities that are permeating Panorama, focusing on leisure and tourism.

The new territorial dynamics in Panorama were motivated by the construction of the Engenheiro Sergio Motta Hydroelectric Power Station in the district of Porto Primavera - SP, which was reflected in Panorama - SP with the formation of the artificial lake on the Parana River. This event had environmental, social and economic impacts. The environmental impacts include the flooding of a vast area which affected the local fauna and flora. Social impacts include the loss of spaces that reflected the history of the municipality and the displacement of people (forced by the formation of the lake) who lived near the river banks, which exposes feelings of identity and belonging. The economic impacts are focused on territorial and structural loss, and the decline of the pottery economy, which was the main economic activity. As a result of the impacts caused by the formation of the lake, Panorama has undergone a number of transformations that have resulted in new territorial formations, including environmental, social and economic aspects.

Because of the impacts, the São Paulo State Energy Company (CESP) has carried out a number of mitigation works, including a spa for leisure activities. With the formation of the lake, the clay deposits were submerged, making it difficult to extract clay, which is the main raw material for the ceramics industry. With the lack of raw materials, this sector, which was Panorama's highlight, was disrupted, leading to a lack of jobs, as the municipality depended on this economy to generate employment.

Faced with the dismantling of the ceramics sector, CESP "sells" the idea that the works carried out, in return for the impacts, lead to social and economic development, as they create new territorial arrangements that can be used to generate new economic activities. The geographic and leisure aspects of Panorama led CESP and the municipality's managers to conceive of "tourism" as a way of generating new jobs, and that this activity could lead to social and economic ascension. Through the new territorial dynamics that have taken hold in Panorama, this work aims to analyze leisure and tourism, since there is a gap between these modalities that will be understood in the course of the work. In order to understand these new territorial dimensions of Panorama, it was necessary to look at the history of the area, contrasting it with the present. The ideas of Santos (1982 and 1994), Santos and Silveira (2003), Cruz (2002), Coriolano (2006), Haesbaert (2004), Saquet (2010), among others, are used as a theoretical basis for this work.

Therefore, the in-depth study of the new territorial dynamics of Panorama, emphasized by tourism and leisure, seeks to contribute to the discussions on tourism and leisure territoriality, showing its main approaches and the manipulation of this activity by territorial managers who, through *marketing,* create "tourist" spaces or imaginary tourism. To this end, this work is structured in three chapters, as follows:

The first chapter presents the general objective, the specific objectives of the research and reflections on fieldwork and collection procedures.

The second chapter focuses on tourism, looking at the historical aspects of this activity in the world and in Brazil in terms of global/local logic. The theoretical conceptualization, analyzing concepts such as: tourism, tourist, tourist attraction, tourist equipment, among others. Also in this chapter, the concept of territory is developed in order to understand the development of tourist territories.

The third chapter consists of a temporal analysis of Panorama, emphasizing its territorialities. This historical structure unfolds in three moments in this work: the before, reflecting on the history of Panorama, since its foundation, highlighting its location and geographical aspects. The history has been carried out between the "old and the new", focusing on urban and economic development and the relationship between society and space. The second section discusses the major engineering works, from the perspective of Milton Santos, and their impact on Panorama, focusing on the construction of the Sergio Motta HPP and the Mario Covas Bridge. Seeing these works as policies that stem from global movements and manifest themselves locally, the third moment of this chapter, considered the "after", focuses on changes in the territory and landscape. The new dynamics of Panorama after the implementation of the engineering systems are discussed, highlighting leisure and "tourism" and providing a theoretical framework for this activity in the municipality. In view of the new territorial configurations that are unfolding, various views of the society that makes up Panorama on the new reality experienced are analyzed.

RESEARCH OBJECTIVES

This research aimed to analyze the production of territory in Panorama, between tourism and leisure, following the implementation of the Sergio Motta HPP. It is located in the district of Porto Primavera. Since the construction of the hydroelectric power station and the formation of the artificial lake on the Parana River, the municipality of Panorama has undergone transformations of various kinds, of which we will emphasize tourism and leisure. In view of the new territorialities that have been materializing in the municipality, it is necessary to understand their impact on society. We should highlight the economic changes that have been brought about by the impacts and insertion of a new economy, which is based on the ideas of CESP and Panorama's representatives, in an attempt to "sell" the idea of development. From the point of view of CESP representatives, Panorama is a "tourist" town and, based on this idealization, the research will look at tourism concepts and theories to find out what kind of activity is carried out in the town. The question is whether Panorama is a tourist town or whether it has just been structured to emphasize leisure, attracting, according to Cruz (2001), day tourists or tourists who use fishing tourism.

Thus, the main focus of this work/research is to understand the new territorialities of Panorama through a space-time analysis, dimensioning Panorama between tourism and leisure. Among the specific objectives that guided the research, we can mention the analysis of the transformations that have taken place in Panorama, starting with the implementation of the Sergio Motta hydroelectric plant in its socio-economic and environmental aspects, emphasizing tourism and leisure and the political aspects that outline the territories from a global/local perspective; and the understanding of leisure and tourism practiced in Panorama from a social perspective. The objectives of the research are also to understand the major works, i.e. engineering systems, which provide new qualities to the territory, and the relationship between society and space; to analyze the historical context, making a counterpoint with the present, understanding the forces that give movement to the territory; understand the territorial dynamics between tourism and leisure in Panorama, which is considered a border region between the states of Mato Grosso do Sul and São Paulo, highlighting factors that give new designs to the "tourist" territory and regional development; to analyze the relationship between the population and the "new" economic activity, addressing its insertion into the activity and planning, as well as analyzing the discourse of representatives regarding the new dynamics inserted into the territory, putting tourism and leisure on the agenda, contrasting with the impacts caused by CESP, and finally, to understand the transformations of the territory by tourism and leisure, through the influence of investments and actions by the public and private sectors.

Therefore, in order to interpret the creation and production of "tourist" territories and the expansion of leisure in Panorama following the implementation of the Sergio Motta hydroelectric power plant, a specific bibliographical review was carried out, in order to understand the concepts and themes that will be present in the discussion of this work, a historical-documentary search and photographic records of the municipality of Panorama, which were necessary to understand the transformations of the different moments of the municipality, emphasizing the "tourist" activity and the expansion of leisure. *On-site* visits and open-ended interviews with the municipal tourism office and local authorities (Secretary of Tourism).

Tourism, Mayor), emphasizing the planning and insertion of tourism activities in the municipality and their repercussions at regional level.

REFLECTIONS ON FIELDWORK AND COLLECTION PROCEDURES

Fieldwork is of the utmost importance to the geographer, providing him with an integrated analysis of reality and a view beyond the landscape, contributing to a totalizing analysis. In this sense, we can say that fieldwork is a tool of excellence for the geographer, where he can transpose and test theories in his laboratory, which is the geographical space. In this respect, Alentejano (2006) mentions:

Fieldwork therefore represents a moment in the process of producing knowledge that cannot do without theory, otherwise it will become empty of content, incapable of helping to reveal the essence of geographical phenomena. (ALENTEJANO, 2006, p.57)

In this way, theory must precede fieldwork, providing a consistent theoretical-methodological basis for applying it in the field, contributing to the production of knowledge and broadening the geographical lens. Through fieldwork, the researcher finds himself in a new world, which enables him to get in touch with reality.

The production of knowledge implies two aspects, that of the researcher and that of society, because through fieldwork data is collected that increases knowledge about the area studied, which provides personal and professional growth for the researcher, and which may or may not serve the good of society. As mentioned, fieldwork is a tool and has various uses, which can emphasize society or its domination, and through fieldwork it is possible to acquire various data and a complex knowledge of the area studied, using the experience of this space and the theoretical transposition. In this perspective of looking beyond the landscape and understanding the space as a whole, Lourenço (1991) emphasizes:

The immediate question is how to deal with the diversity of landscapes in a way that doesn't stop there but, on the contrary, goes beyond the immediate, the apparent, the empirical. To do this, it is necessary to understand it as an external manifestation (photograph) of a content (society) that defines and elaborates it (LOURENÇO, 1991, p.23).

According to Lourenço (1991), fieldwork must go beyond the limits of the landscape. In this sense, this broad vision becomes a tool that can be used for a variety of reasons, such as its ability to get to know spaces and demystify reality, which can help society or manipulate it.

In order for this tool not to be trivialized and lose its meaning, it is necessary to plan, both theoretically and in terms of fieldwork. The role of the geographer is to carry out an integrated analysis between physical and human geography, in which Suertegaray (2002) mentions the importance of the relationship between the two geographies, since both are interrelated. The magnitude of geotechnologies makes detailed geographical analysis possible, but these tools do not exclude fieldwork, as they represent a further aid to geographers in their work.

From technique to reality, we were able to observe the city of Panorama before applying the questionnaire, showing the multiplicity of spaces in the same area, and the research area is very complex, as we have a new activity and the remains of another. In this sense, this space presents various readings that will depend on the engagement or political position of those who manage the territory and the research. In this sense, the work and research have more than met expectations, broadening horizons and qualifying these practices in both the theoretical and practical spheres.

For this analysis, the research was based on the concept of territory and tourism, which provided a theoretical and methodological contribution to understanding the dynamics installed in Panorama. As for the procedures, photographs from different periods were used for temporal analysis.

Formal and informal open-ended interviews were conducted with the municipal tourism secretary, the mayor, the local community, the president of the artisans' association and traders.

This was one of the main stages of the research, which made it possible to assess the various discourses and interests involved. The fieldwork and interviews were carried out on different dates during 2011/2012. Since tourism and leisure in Panorama suffer from seasonality, this provided different analyses of the flow, circulation of tourists and visitors, and jobs throughout the year. The fieldwork was carried out during periods considered to be high season, from March to October, especially when fishing is allowed, and events such as the Carnival holidays. And in the low season, from November to February, centered on the piracema, the season when fishing is prohibited. Seasonality has a direct impact on the population that lives off tourism and leisure in Panorama, especially the sectors that depend directly on fishing. In this sense, it was important to talk to the population that depends on tourism and leisure. Talking to tourists and visitors was fundamental to understanding the meaning of leisure and tourism in Panorama, pointing out the positive and negative aspects of the activity.

The research and knowledge of the study area provided new visions that reflected in the deepening of the discussion and new analyses, incorporating the global relations that manifest themselves in the local and the valorization of tourism to the detriment of society.

In this respect, there has been a great deal of commitment to demystifying the territorial representations, formed by the idealizations of the representatives of the municipality and CESP, in the construction of a "tourist" city. Through the mitigation works carried out in Panorama, the urban and natural elements have been valorized, and through marketing, the managers of the territory have been commercializing the spaces, creating the image of a tourist town that focuses on fishing. This trend is related to the decline of other activities and, in this sense, the municipality faces this position due to the lack of choice and imposition in the face of the impacts caused by CESP. In this way, Panorama finds itself between "tourism", created by marketing, and leisure, which has been around since the town was formed.

2. THEORETICAL BACKGROUND TOURISM

Tourism has been gaining prominence in today's globalized world, where interactions and flows are denser and society is shaping itself according to the dynamics established and, in this context, tourism causes changes and new formations in the relationships between social agents and space. In this sense, Rodrigues (1996) mentions:

In a globalized world, tourism comes in many forms, in various stages of evolution, which can occur synchronously in the same country, on a regional or local scale. It is expanding globally, sparing no territory [...]. It is certainly a complex phenomenon, designated by different expressions: a social institution, a social practice, a pioneering front, a civilizing process, a system of values, a lifestyle - a producer, consumer and organizer of spaces - an "industry", a trade, an interwoven and improved network of services. (RODRIGUES, 1996, p.17).

This activity is not exclusive to the globalized world, or as we think, the result of the modern economy. Tourism arises from society's curiosity, pleasure and leisure, and there are various reasons that can turn a simple trip into tourism. Researchers have debated whether or not travel can be tourism, because since ancient civilizations, society's movements have been motivated by wars, the search for better places or even to discover new lands and continental masses. In this sense we can't characterize them as tourist trips, since to characterize a tourist trip we must have the element of pleasure, curiosity and leisure. In the discussion on the temporality of tourism and travel, Coriolano (2006) argues that:

Unlike tourism, travel has existed since the beginning of human existence and took place for a variety of reasons: the need to survive, wars, sacred rites and the search for health. Of course, in ancient times, these journeys didn't have the connotations of modern tourist travel (CORIOLANO, 2006, p.22).

In view of the theoretical notes on tourism as an activity, and travel, it can be seen that primitive displacements are not characterized as tourism, because they do not include leisure. According to Coriolano (2006),

Tourism is a phenomenon of modern times, an invention of capitalism, and is therefore relatively recent. It arose when man discovered the pleasure of traveling, not just out of necessity and obligation, but because it was something pleasurable, a form of enjoyment, until it became a commodity as an object of desire and happiness (CORIOLANO, 2006, p.21).

The first trips considered to be tourist trips were the "Grand Tour", organized in the 18th century. It was educational tourism and began before leisure tourism with the study trips of the young English nobility. With regard to the Grand Tour, Andrade (2000) emphasizes that:

The Grand Tour, under the imposing and respectable title of "study trip", took on the value of a diploma that conferred significant social status, although - in reality the program was based on great tours of excellent quality and full of pleasurable attractions (...). The English, important and wealthy, considered only those who had their education or professional training crowned by a Grand Tour through Europe to be detainees of culture (...) (ANDRADE, 2004, p.9).

In this sense, Andrade (2004) mentions that in the 18th and 19th centuries, noble families sent their children to study in the great cultural centers of Europe, accompanied by their competent and illustrious preceptors. These trips were labeled "Grand Tour".

Through the history of tourism and its first initiatives and meanings, we can see its constant transformation and its new meaning in the contemporary world. Traveling for leisure, in search of knowledge, or even for pleasure goes back a long way, but nowadays travel and tourism have been boosted by the economy, coming to be thought of as an economic asset and a social practice, which allows us a deeper and more complex analysis of the subject. Through travel and man's need to get away from his daily life, tourism emerges as an economic mechanism and social practice that acts on the territory, creating new dynamics. From this perspective, according to Coriolano (2006):

Tourism studied as a phenomenon involves different approaches, and in any of them it is necessary to highlight the role of space in restructuring the economic system and the territory itself. In this activity, places, markets, people, jobs, work and policies interact as the driving force behind regional development (CORIOLANO, 2006, p.40).

Tourism gained momentum and massification with the industrial revolution, a period of great importance in which society acquired new behaviors and new needs. Mass tourism is gaining importance nowadays, and in this scenario its conceptualization is confused with popular tourism. In this sense, Cruz (2001) points out that:

...it must be recognized that "mass tourism" does not mean "tourism of the masses", for the simple reason that the masses do not do tourism. Mass tourism is a way of organizing tourism that involves agency activity as well as the interconnection between agency, transport and accommodation, in order to lower travel costs and consequently allow a large number of people to travel. It should be remembered, however, that this number of people traveling is very far from corresponding to the total world population and therefore very far from corresponding to the mass of the planet's population. (CRUZ, 2001, p.6)

With the popularization of tourism, it has become one of the necessities of the modern world, boosting the consumption of goods and spaces, as Coriolano (2006) points out:

In industrial society, the activity became a mass phenomenon, standardized and able to serve a greater number of middle-class people. It began to generate profits and foreign exchange, sometimes at the expense of degradation, spatial de-characterization and social discrimination (CORIOLANO, 2006, p.32).

With work and the urban world, a routine is created that is broken with vacations, which are a time for leisure and an opportunity to get away from the daily grind in order to re-establish one's "energies" in terms of the man power and work of the system. In this sense, we can relate the expansion of tourism to the capitalist perspective, whose very system creates a dependency on rest and, consequently, on leisure and tourism. To demonstrate these aspects of modern life, Coriolano (2006) points out:

In modern life, both the feeling that work is stressful and the frequent rush of the city have made people prioritize the need for leisure and the search for happiness outside of everyday life. As many people see it, finding happiness means leaving the real world of work, made possible by a tourist trip. Many believe that happiness is easily found outside of everyday life (CORIOLANO, 2006, p. 23).

Tourism is valued with capitalism, which creates the need for leisure in society, with vacations or moments of leisure in which society can enjoy and get away from its daily routine. In this vein, Bertoncello (1996, p. 209) points out that "The habit of traveling and tourism is nowadays firmly incorporated into consumer society as a necessity, and its satisfaction gives rise to the development of specific activities of great importance".

From this perspective, capitalism begins to commercialize spaces, creating artificial spaces, moving new territorialities, turning history, people and nature into commodities and commercializing them. In the globalized, industrial and urban world, new forms of tourism have emerged, such as adventure tourism, ecotourism and rural tourism, which are part of the vision of escaping the urban world and pollution. These are some of the parallels between tourism and its potential and the connotations it has received over time. These connotations are also mentioned by Coriolano (2006):

Leisure, being a vital need, is enjoyed by everyone and can be done without spending money. Tourism is leisure transformed into merchandise. It is sophisticated leisure that requires travel, and is therefore an invention of consumer society, responding to the needs not directly of man, but of capital. (CORIOLANO, 2006, p.217)

Tourism has many gaps in its body of theory, as it depends on different areas and this leads to different views on the subject. Tourism depends on economics, geography, administration and other areas that together make up tourism. But in the field of research, this interdisciplinarity creates various conceptual debates, making it difficult to systematize theory and methodology. One of the debates starts with the name itself: What is tourism? The debate is present in academia and takes on different views depending on the professional who is analyzing it.

There is a need for conceptual and methodological systematization on the subject of tourism, since this activity is one of the engines of the world economy and a major shaper of space, requiring planning and policies to monitor and enhance it. In this sense, tourism needs greater attention in the field of research so that it supports social development and not its emptying.

With the capitalist approach to tourism, this activity has become a tool for exploiting space, turning it into a commodity as a whole: the history of the place, the people, the architecture, everything becomes a commodity in tourism. In this sense, it is important to systematize the theory so that it reaches the reality and potential of each tourist area. But it's worth pointing out that tourism is often seen in a technical way and with concepts that support capitalism in its spatial appropriation. For this reason, geography has become a tool for managing space and demystifying the process of commercialization of space by tourism. In the wave of globalization, whose networks and flows are more accentuated, there is a greater proportion of the flow of tourists in the world, changing the meaning of spaces. Every so-called tourist area must be planned so that it doesn't alter its natural and social characteristics, but only leads to the socio-economic development of those affected by tourist activity.

Tourism is associated with movement in various forms and in various contexts, which allows us to make various analyses of this activity, since the flow and the goods - "spaces" - have changed in the course of tourist activity and, in this sense, a conceptualization of tourism theories has become necessary.

With the transformations in society that directly influence tourism, its definition (tourism) becomes important for academic use and the planning of this activity. For the UNWTO (2001), tourism comprises the activities that people carry out during their trips and stays in places other than their usual surroundings, for a consecutive period of less than a year, for leisure, business and other purposes. Cruz (2003, p.5) understands that tourism "is first and foremost a social practice involving the movement of people across the territory and which has geographical space as its main object of consumption". In the view of Coriolano (2006), "tourism is a form of leisure and entertainment that requires travel, the movement of people, consumption, free time and the use of equipment, however minimal, such as transportation, hotels, inns and restaurants".

According to Andrade (2004, p.38), "tourism is the complex of activities and services related to travel, transportation, accommodation, food, the circulation of typical products, activities related to cultural movements, leisure and entertainment".

In view of the definitions presented, the UNWTO differs with a more technical approach, opening up the possibility of other interpretations of the concept of tourism. The conceptualizations and theoretical/methodological concerns about tourism are recent, focusing on studies in different areas, which makes debate and a theoretical proposal difficult. For the UNWTO there is a great deal of tourist segmentation, which makes it clear that every trip is considered tourist, even business trips, which opens up a discussion about what leisure is and whether tourism should be related only to leisure and pleasure. At the same time as this conceptualization opens up the possibility of classifying all trips as tourist, it is limited to overnight stays and length of stay, which excludes short-distance tourism.

From a less technical point of view, we can look at the use of territory, because the moment tourists move from one territory to another, it can be considered tourism, since they will be out of their reality, experiencing another culture, habits, consuming both merchandise and spaces commercialized by capitalist ethics. Tourism can be conceptualized through the consumption of territories, since these are equipped with culture, identity and economy. The moment a person leaves their territory, they come into contact with the other and another territory, which can be considered tourism, analyzed from the point of view that they are practicing leisure, and this also gives them pleasure and, on the capitalist side, they are consuming what this territory offers, moving the economic dynamics of this activity.

The above statements show us the shortcomings of each and their complementarity, since the statements are opposites: one has a technical character, aimed at the economic side, while the other presents a human character and use of the territory, which can also be interpreted from an economic point of view. The UNWTO's definition is validated since, through it, we can analyze that even on a business trip, the tourist can be enjoying and consuming the territories and their tourist attractions, but is limited by the length of stay, which can be complemented with the second view, that it is enough to move away from their daily lives and consume the territory.

Another important concept for understanding the dynamics of tourism is the tourist. This concept has acquired various connotations over time, and is the main asset of this activity. In 1954, the United Nations (UN) defined tourists as:

Any person without distinction as to race, sex, language or religion who enters the territory of a city other than that in which they are habitually resident and stays there for a minimum of 24 hours and a maximum of six months, during a 12-month period, for the purposes of tourism, recreation, sport, health, family reasons, study, religious pilgrimage or business, but not for the purpose of immigration.

According to the Ministry of Tourism (2007), a tourist is "a traveler who goes to one or more places other than their usual place of residence and stays there for more than twenty-four hours, but with the intention of returning. In addition, they do not participate in the labor market at the destination". In contrast to the concept of tourist, the concept of excursionist or day tripper is centered on the tourist, who has the same characteristics as the tourist, except for the "overnight stay", which is the difference between the concepts. Andrade (2004) explains the concepts:

The term excursionist or day tripper is used to describe those who travel and stay for less than twenty-four hours in a place other than that of their fixed or habitual residence, for the same purposes as tourists, but without staying overnight in the place visited. (ANDRADE, 2004, p.44)

In order for tourism to function, you need tourist equipment, which refers to infrastructure and services. According to the Ministry of Tourism (2007), tourist equipment and services are the "set of services, buildings and facilities that are indispensable to the development of tourist activity and that exist as a result of it. They include services and equipment for accommodation, food, agencies, transport, events, leisure, etc."

Tourist attractions are natural spaces or spaces created to attract visitors and tourists. *Marketing* plays a key role in enhancing these spaces and turning them into tourist attractions. In relation to tourist attractions, there are vast definitions that serve the hidden interests of tourism. For Beni (2008), tourist attractions can be considered one of the main elements of the tourist product, after all, it is with the aim of visiting them that tourists travel to a particular location, and each attraction has its own particularity. In the view of EMBRATUR (1992), tourist attractions represent places, objects or events of interest to tourism, to which can be added the habits and customs of peoples, as intangible elements relevant to the field of tourism. Attractions can be classified according to the natural or humanized potential of a tourist territory, as shown in Table 1, drawn up by the Ministry of Tourism (2007):

Table 1: Category of Tourist Attractions

Categorias	Definições	Exemplos
Atrativos naturais	Elementos da natureza que, ao serem utilizados para fins turísticos, passam a atrair fluxos turísticos.	Montanhas, rios, ilhas, praias, dunas, cavernas, cachoeiras, clima, fauna, flora etc.
Atrativos culturais	Elementos da cultura que, ao serem utilizados para fins turísticos, passam a atrair fluxo turístico. São os bens e valores culturais de natureza material e imaterial produzidos pelo homem e apropriados pelo turismo, da pré-história à época atual, como testemunhos de uma cultura.	Artesanato, gastronomia, museus, festas e celebrações, manifestações artísticas etc.
Atividades econômicas	Atividades produtivas capazes de motivar a visitação turística e propiciar a utilização de serviços e equipamentos turísticos.	Fabricação de cristais, agropecuária, extrativismo etc.
Categorias	Definições	Exemplos
Realizações técnicas, científicas e artísticas	Obras, instalações, organizações, atividades de pesquisa de qualquer época que, por suas características, são capazes de motivar o interesse do turista e, com isso, propiciar a utilização de serviços e equipamentos turísticos.	Museus naturais, observatórios, aquários etc.
Eventos programados	Eventos que concentram pessoas para tratar ou debater assuntos de interesse comum e negociar ou expor produtos e serviços; podem ser de natureza comercial, profissional, técnica, científica, cultural, política, religiosa, turística, entre outras, com datas e locais previamente estabelecidos. Esses eventos propiciam a utilização de serviços e equipamentos turísticos.	Feiras, congressos, seminários etc.

Source: Ministry of Tourism: Roteiros do Brasil. Tourism Regionalization Programme (2007).

0 concept of attraction brings us back to the commodification of spaces and all their crystallized materiality. In this sense, traditions and cultures become tourist products that can be incorporated into itineraries. In the marketing vision of tourist spaces and in the search for the new or escape from everyday life, Coriolano (2006) states that:

0tourism, however, requires specialized places, and is allocated to the most beautiful and well-kept ones; this is a request of the activity itself, considering that tourists are buying an expensive service and, above all, fleeing the problems in the sending poles, not wanting to find them in the receiving ones. (CORIOLANO, 2006, p. 223).

The concepts presented are twofold: the technical nature of the concepts is aimed at the logic of the market, in an attempt to reinforce the sale of spaces and manipulate consumption, which is at a voracious stage, guided by globalization. And the academic character aims to broaden the view of this phenomenon that consumes spaces, thinking about their social integration with this economy that is becoming more representative every day. With technology, spaces undergo sudden changes in order to serve capital and its expansion. Tourism, seen as a segment of capital, takes advantage of both the techniques and the very system that drives it. In the creation of spaces and new configurations of tourism as a human need, Santos (2000) shows how technology has produced another "tourist place"

We've gone from localized artisanal leisure to globalized industrial leisure, from leisure embedded in society to automated leisure. This automation has turned leisure into an industry, an operation in which the various parts are a system, be it tourism, sport or any kind of entertainment. The revolutions in transport, sound, image and telecommunications have integrated places and their forms of representation, such as image, sound and message, generating interactivity, with greater possibilities for relationships and mutual responses between residents and tourists. However, the great contradiction lies in the fact that instead of having a truer knowledge of places, with their variety of combinations and choices, with the biased commercial production of images of the world and images of people, we are under the permanent threat of manipulation. The stereotyped image threatens to replace the taste for fantasy and discovery (SANTOS, 2000, p.32).

TOURISM IN BRAZIL

Tourism, as you can see, is not an activity restricted to the successor world. When we analyze tourism on a global scale, we see a large flow of people, especially in the so-called developed countries. This international flow is related to various factors such as tourist infrastructure, attractions, planning and, above all, marketing.

Data presented by the UNWTO shows the importance and growth of this sector on a global scale. Between 1999 and 2010, the international flow of tourists in the world grew by 49%, emphasizing the magnitude of this sector, making it one of the main economies in the world. In view of the growth of this activity, Moesch (2000) mentions that:

Tourism was born and developed with capitalism. With every advance in capitalism, there is an advance in tourism. From the 1960s onwards, tourism exploded as a leisure activity, involving millions of people and becoming an economic phenomenon, with a guaranteed place in the international financial world. [...] (MOESCH, 2000, p.9)

Brazil has great potential to add value to its territory and become a major tourist power. According to the UNWTO, in 2011 Brazil attracted 0.**5%** of the flow of international tourists, which indicates that this segment is tending to expand, becoming one of the country's main economic activities. In 2011, around five million foreign tourists visited Brazil, spending around R$12.5 billion, according to data from the *World Travel & Tourism Council* (WTTC).

The increase in average income, the availability of credit and the lower cost of airline tickets - an average reduction of 34% in the last 10 years (data from the Ministry of Tourism) - have allowed the C, DE and E classes to travel more within our country. According to the UNWTO, tourism grew by 4.4% worldwide in 2011 and by 11.7% in Brazil. The data presented shows that Brazil is not yet among the main tourist hubs internationally. But it has the potential to expand and become one of the world's main tourist destinations, highlighting its potential in the natural environment, which is valued in the modern urban world. Focusing on the production of spaces and relating them to Brazil's tourist potential, Carlos (1996) emphasizes that:

More and more space is produced by new sectors of economic activity such as tourism, and so beaches, mountains and countryside enter the circuit of exchange, appropriated privately as leisure areas for those who can make use of them (CARLOS, 1996, p. 25).

On an international scale, the data presented shows the discrepancy and economic importance of tourism on a global scale, in which underdeveloped countries are less dynamic due to the lack of infrastructure and policies that guide tourism in these countries. Among the characteristics of tourism in underdeveloped countries is the lack of planning, professionalism, investment and tourist mentality. Part of the tourism practiced or sold in poor countries can be characterized as amateurish. Amateurism in this activity leads to a lack of commitment to social and environmental issues, since it appropriates and modifies environments and must be thought out and planned for by everyone involved.

Tourism as a social practice must go beyond the meaning of leisure and must be carried out in a planned way with qualified people so that it leads to economic development and the social upliftment of the community.

In Brazil, tourism has the characteristics mentioned above, practiced for the most part without social responsibility and commitment, without planning, compromising socio-economic development. Brazil has great tourism potential, be it for its culture, its history, but above all for its natural potential, which is being valued in today's world. Given its potential, tourism can be a way of boosting the economy and development, especially on a local level. However, in order to bring about development, it is necessary to think from various perspectives, especially in terms of the present society and the tourists, not just looking at the economic potential. Tourism, viewed only from the economic point of view and without planning, can be the way out for socio-economic growth, while the misuse of space and its deterioration can bring tourism to an end, because tourism depends on natural or modified space to carry out this activity. In this respect, Rizzo (2010) emphasizes that tourism in Brazil

...it's only not bigger for a few reasons, among which are the lack of complete infrastructure for tourist services; the unsatisfactory level of security; the lack of adequate infrastructure for tourists and the lack of investment in publicizing Brazil abroad and internally. (RIZZO, 2010, p. 202)

Tourism in Brazil has gone through several phases, but its crystallization began in the 1960s, with structural and social changes. With the expression of urbanization in the 1950s, a new mentality embraced Brazilian society and new patterns of consumption, said to be from the "modern" world, including leisure and travel. Capitalism as the creator of the need for leisure, represented by vacations, boosted leisure and tourism worldwide. In Brazil, with the development of industrialization and the consolidation of the CLT (Consolidation of Labour Laws) with rights such as vacations, among others, creating the conditions for a new market that explores leisure and tourism. Among the events that gave Brazil a tourist outlook was the World Cup held in Rio de Janeiro in 1950, which helped to publicize Brazil's tourist potential abroad, promoting its culture and natural attractions, and attracting foreign tourists to the country.

In the 1970s, tourism gained importance worldwide and was inserted into the political context as an economic activity capable of solving the problems of the most stagnant localities. In the 1990s, the world entered a new phase, called globalization, where technological means became more present in the world, shaping space and accelerating social and economic relations, making space suitable for receiving tourist dynamics, turning the local into global spaces, through marketing and consumption. In this phase (globalization) tourism has acquired new connotations and relationships, in this sense Rizzo (2010) mentions that:

In the sense of globalization, competition for tourists today is also globalized and internationalized, which requires companies, attractions, accommodation facilities and municipalities to constantly and quickly update information for tourists. Tourism is increasingly demanding professionalization and is no longer an amateur activity (RIZZO, 2010, p.206).

However, Brazil has a shortage of studies and research focusing on the tourism market, which makes it difficult to understand the dynamics of this activity and its consequences in places where tourism is developed spontaneously, without any kind of planning or analysis of the impacts that the activity could have on communities.

With regard to the formalization of tourism and its policies in Brazil, in 1958, the Federal Government created the Brazilian Tourism Commission, which was abolished in 1961, and in its place, in 1962, the Tourism and Events Division of the Ministry of Industry and Commerce was created. Faced with the growth of the activity in Brazil and seen as a local development policy, in 1996 the document instituting the National Tourism Policy was created, but some development programs had already been created in 1995, such as the PNMT (National Tourism Municipalization Program), aimed at both decentralizing the development of the activity and identifying Brazilian municipalities with tourism potential. As Rizzo (2010) mentions:

With regard to the institutionalization of tourism in Brazil, there was an important advance in the first government of President Fernando Henrique Cardoso: in 1995 the National Program for the Municipalization of Tourism (PNMT) was implemented, which was designed to give municipalities more autonomy in tourism planning, since each municipality has its own characteristics such as environmental and cultural heritage and even the traditions and customs of each population. (RIZZO, 2010, p.195)

With a view to structuring and planning tourism activity, the Federal Government created the Ministry of Tourism in 2003, drawing up the National Tourism Plan (PNT), with the aim of increasing activity in Brazil by creating new tourist hubs, guiding policies and resources in international dissemination and promotion. According to the PNT (2007/2010):

Tourism in Brazil will take into account regional diversity, (...) the creation of employment and occupation, the generation and distribution of income, the reduction of social and regional inequalities, the promotion of equal opportunities, respect for the environment, the protection of historical and cultural heritage and the generation of foreign currency (...) (BRASIL - PNT 2007/2010)

In view of the policies instituted to support and guarantee the development of tourism in Brazil, new policies have been incorporated and modified to meet the needs and peculiarities of each region and municipality, and Rizzo (2010) emphasizes that:

Returning to the institutional field, another important step came during the government of President Luis Inacio da Silva, when the PNMT was revised and replaced by the Tourism Regionalization Program - Roteiros do Brasil (PRT), which began to plan and execute tourism in the regional area, but without extinguishing the responsibilities of the municipalities. The objectives of the PNMT were to improve the living conditions of the receiving communities, such as generating jobs and improving income distribution (RIZZO, 2010, p. 200).

With the creation of the Ministry of Tourism and the development of its policies, this activity gained territorial representation and the possibility of becoming a social asset, leading to socio-economic development. From this perspective, PNT (2006) points out that:

The regionalization of tourism, implemented by the Tourism Regionalization Program - Roteiros do Brasil, launched in April 2004, proposes the structuring, planning and diversification of the country's tourism offer and is the territorial reference point for the National Tourism Plan (PNT, 2006).

Through this scenario it is possible to see how important this activity is for the country and that, through planning, it can lead to the development of the community affected. The creation of the Ministry of Tourism opens up the possibility of planning and guiding policies aimed at this sector, re-evaluating both parties that tourism affects: the tourist and the local population.

TOURISM AND ITS LOCAL DEVELOPMENT

Tourism has become a sector of great magnitude and currently stands out as one of the main economic activities. However, this activity is not only attracting attention for its economic potential, but also for its ability to transform spaces, create new dynamics, among other functions that are attracting the attention of researchers from various fields, ranging from economic studies to sustainability. In this respect, Caracristi (1998) points out that:

It is an activity that, especially in recent decades, has had a decisive influence on the space in which we live, producing significant economic, social and environmental transformations: The tourism industry is currently the fastest growing activity in the world economy (CARACRISTI, 1998, p.407).

As mentioned above, tourism has emerged as one of the main cogs in the wheel of capitalist accumulation, inserting new strategies into space that are commanded by various sectors of society, in this sense Coriolano (2003) emphasizes:

Tourism is one of the newest forms of the accumulation process, which has been producing new geographical configurations and materializing space in a contradictory way, through the actions of the state, companies, residents and tourists (CORIOLANO, 2003, p. 42).

In the perspectives presented on the product and consumption of tourism, its main commodity is space, appropriating natural elements, making them marketable or economically viable for exploitation. Tourism has a great capacity to transform spaces and the agents that are part of them, such as the population, commerce and others. In this line of thought, Coriolano (2003) mentions:

By transforming space into a commodity, capital gives rise to new economic activities, such as the economic branch of leisure activities and tourism and leisure. Tourism causes profound socio-spatial change, redefines spatial singularities and reorients uses. (CORIOLANO, 2003, p.32)

Tourism can be seen in two directions, one by the possibility of bringing development to the local population, increasing jobs, improving the infrastructure that is used by the local community and tourists, the other from the perspective of offering excellent services to tourists and keeping the activity active in a sustainable way, valuing cultural representations and the organization of society. For many Brazilian municipalities, tourism is the only source of income, especially small towns that are not part of the industrialization process or have not specialized, and rely on tourist attractions, whether cultural or natural, making it the population's only source of income. Faced with the possibility of tourism raising the quality of the community involved, Veloso (2003) mentions that:

The various types of tourism practiced around the world make this activity a great option for development. Each location needs to define which type or types of tourism its characteristics fit into, according to the region's potential. This definition is important not only to provide visitors with information about the types of tourism that the locality offers, but also to guide those who want to invest in the sector. (VELOSO, 2003, p.13)

In order to carry out a tourist activity, it is necessary to take into account various presuppositions, such as planning, the inclusion of the population, tourist knowledge and professionalism, so that the activity provides local development and that it is practiced in a "sustainable" way, so that tourism does not deteriorate the infrastructure, attractions and equipment, which could lead to the decline of this activity. In order for this activity to last, it must be properly planned, from the reception of tourists, according to the type of tourism practiced, to the appropriate infrastructure, among other elements that are important for the development of the activity. In relation to tourism planning, Trigo (2001) states that:

Tourism planning is the process that aims to order human actions on a tourist site, as well as directing the construction of equipment and facilities in an appropriate manner, avoiding negative effects on resources that could destroy or affect their attractiveness. (TRIGO, 2001, p.67).

Planning has become an essential tool for tourist activity, providing orderly investment and maintenance of tourist infrastructures and equipment, avoiding their deterioration, which could lead to the end of the activity.

Among other aspects relevant to tourism are the positive and negative impacts, of which the environmental, socio-cultural and economic ones stand out. Tourism is characterized by the appropriation of spaces, turning them into commodities that are consumed by tourists and which can lead to their degradation, requiring monitoring and planning, preventing negative impacts from leading to the decline of this activity, which can be irreversible. On the other hand, this activity, carried out in a planned manner, can lead to development in various aspects, such as social, environmental, structural and economic. With regard to the positive and negative impacts caused by tourism, below is a table drawn up by the Ministry of Tourism (2007), segmented into environmental, socio-cultural and economic impacts.

The variables presented (Table 2) in relation to the positive and negative impacts of tourist activity show the complexity of their analysis and monitoring, not depending only on tourism, but on a range of knowledge that requires interdisciplinary work. In order to make tourism and its monitoring efficient, there needs to be coordination between the different segments of government, business, civil society and educational institutions in order to monitor these indicators.

Reading Table 2, it is possible to make various comparisons with Panorama, since this municipality has suffered different impacts from the expansion of leisure and "tourism". In environmental terms, Panorama has been impacted by the formation of the artificial lake, which has led to a loss of land, and by the poor occupation of the land represented by allotments on the banks of the Parana River. From an economic point of view, leisure and "tourism" suffer from seasonal dependence on fishing. With the increase in second homes, there is real estate speculation, which makes life difficult for residents and increases the cost of living. These aspects will be developed in more depth in the next chapter.

Table 2: Positive and negative aspects of the impacts of tourism.

Impactos	Aspectos positivos	Aspectos negativos
Ambientais	• Melhoria dos padrões de uso do solo urbano e rural na região turística. • Manutenção das áreas verdes protegidas. • Aumento das atividades ligadas à educação ambiental. • Melhoria da coleta e destinação do lixo e outros resíduos sólidos. • Redução da poluição ambiental. • Manutenção da qualidade da água. • Melhoria da qualidade do esgotamento sanitário.	• Má utilização do solo e dos recursos naturais. • Ocupação desordenada do solo. • Desenvolvimento desordenado do turismo que venha a provocar degradação ambiental. • Aumento da poluição geral e do lixo produzido por excesso de carga ou saturação da região. • Poluição sonora, poluição visual causada pela propaganda.
Socio-culturais	• Consolidação da identidade cultural com resgate e valorização de atividades típicas da região (danças, música, folclore, artesanato, gastronomia etc.). • Aumento de ações voltadas para o resgate e preservação do patrimônio histórico e cultural (visitas a museus, monumentos etc.).	• Mudanças negativas nos hábitos e padrões culturais e de consumo (alcoolismo, consumo de drogas, prostituição etc.). • Perda da identidade cultural pela influência externa. • Ampliação das desigualdades sociais.
Econômicos	• Diversificação e ampliação das atividades econômicas na região. • Aumento do fluxo e da circulação de dinheiro. • Aumento dos postos de trabalho, principalmente aqueles voltados às atividades da comunidade local. • Aumento e distribuição da renda média da comunidade local. • Inclusão socioeconômica dos segmentos da cadeia produtiva do turismo. • Aumento da demanda por produtos agrícolas locais. • Aumento do consumo de bens e serviços em geral pelas comunidades. • Aumento da competitividade dos produtos gerados no setor. • Contribuição do turismo para o equilíbrio da balança de pagamento.	• Aumento da dependência local e regional da atividade turística em detrimento de outras atividades produtivas. • Sazonalidade da demanda turística, propiciando períodos de recessão econômica. • Aumento do custo de vida e especulação imobiliária. • Ampliação das desigualdades econômicas.

Source: Ministry of Tourism: Roteiros do Brasil. Tourism Regionalization Program (2007).

APPROACHES TO THE CONCEPT OF TERRITORY

The concept of territory has various interpretations as it is used in different areas of knowledge such as biology, economics, anthropology and also geography, but with different configurations. Given the ramifications of the concept of territory, it is necessary to analyze and conceptualize it. In conceptual terms, Haesbaert (2004) explains that:

Etymologically, the word territory, territorium in Latin, is derived directly from the Latin word terra and was used by the Roman legal system within the so-called jus terrendi [...] as a piece of land appropriated within the limits of a given political-administrative jurisdiction [...] (HAESBAERT, 2004, p.43).

For geography, territory has various approaches, which makes it a complex concept with different conceptions within the geographical science itself. According to Saquet (2010):

Territory means nature and society; economy, politics and culture; idea and material; identities and representations; appropriation, domination and control; discontinuities; connections and networks; domination and subordination; environmental degradation and protection; land, spatial forms and power relations; diversity and unity. This means the existence of interactions in and of the process of territorialization, which involve and are involved by similar and different social processes, at the same or different times and places, centred on the paradoxical combination of discontinuities, inequalities, differences and common traits (SAQUET, 2010, p. 24).

Given the complexity and use of the concept of territory, Saquet (2010) presents an approach, emphasizing that:

Territory can be thought of as a text in a context, as a Place articulated to Places, by multiple relationships, economic, political and cultural; it is movement and unity between being and nothingness, (i) materially. It is relaxed and reproduced in a single process. There are subjects and, at the same time, the transformation of being into its other being, which contains it. One is in the other, in the same movement of the formation of the territory. (SAQUET, 2010, p.163)

Saquet (2010, p.129) understands territory "first and foremost as a space for organization and struggle, for the experience of citizenship and the participatory nature of managing what is different and unequal".

Saquet's exposition opens up the importance of this concept and its variation, which can be interpreted from the imaginary to the real, from economic to cultural relations. In this sense, it is a broad methodological concept with diverse applications. Still on the conceptualization of territory, Moraes (2005) defines it as:

An earthly materiality that houses a country's natural heritage, its production structures and the spaces where society (in the broad sense) reproduces itself. This is where the sources and stocks of natural resources available to a given society are allocated, as well as the existing environmental resources. It is also where the spatial forms created by society over time accumulate (produced space). These forms are added to the soil, becoming territorial structures, conditions for production and reproduction at each juncture (MORAES, 2005, p. 140).

The territory defined by Santos and Silveira (2003) focuses on the "used territory", which is organized and constructed through the materialities that crystallize in space over time. In this way, Santos and Silveira (2003, p.247) argue that, "when we want to define any piece of territory, we must take into account the interdependence and inseparability between materiality, which includes nature, and use, which includes human action, i.e. work and politics".

With the advent of globalization, the territory has acquired new ways of organizing itself to serve capital, thus changing the relationship between space and society. But within a territory that has globalizing characteristics, the traditional becomes the differential that opposes the market and standardizing logic. Faced with the opposing forces in the territory (local versus global logic) that crystallize in the territory, Santos (1994) emphasizes that:

Just as everything was not, so to speak, a "nationalized" territory, today everything is not strictly "transnationalized". Even in places where the vectors of globalization are more operative and effective, the inhabited territory creates new synergies and ends up imposing a revenge on the world. Its active role makes us think of the beginning of history, even if nothing is the same as before (SANTOS, 1994, p. 15).

Another contribution to understanding the concept of territory comes from Raffestin (1993), who mentions its formation, in which space precedes territory, since territory acquires its quality through the power relations established in space, in this sense power becomes an essential element in the construction of territory. In this sense, power mediates social relations, which crystallize in space, with society constructing the territory. In the words of Raffestin (1993):

It is essential to understand that space is prior to territory. Territory is formed from space, it is the result of an action carried out by a syntagmatic actor (actor carrying out a program) at any level. By appropriating a space, concretely or abstractly [...] the actor "territorializes" the space (RAFFESTIN, 1993, p. 143).

As for power, this approach is necessary in order to broaden our view of the concept of territory and the relationships that are established between the different levels, which can be political, social and economic, among others. Transposing this to the tourism sector, we can talk about the organization of this territory through the relationships between tourists and the local community, tour guides, tourism agencies and attractions, thus configuring a tourist territory through the power relations that are established between the most distinct segments.

TERRITORY AND TOURISM

The understanding of tourist territorialities is not only explained by relations of power, because through globalization, the world is taking on new configurations and economic, political and cultural activities are undergoing restructuring and differentiating between the "new and the old". In terms of space-time, Santos (1982) emphasizes that:

Everything that exists articulates the present and the past by the very fact of its existence. For the same reason, it also articulates the present and the future. In this way, an isolated spatial approach or an isolated temporal approach are both insufficient. To understand any situation, we need a spatio-temporal approach (SANTOS, 1982, p. 205).

Tourism is structured through globalization, taking advantage of fluidity, technology and the characteristics of the modern world, turning leisure, public spaces and business time into a valuable commodity. In this perspective, capital takes advantage of leisure time to guarantee the flow of tourist activity. Rodrigues (2001) considers:

[...] You can spend your **free time** without doing anything. In this case, the time spent is pure **leisure** time, i.e. time for contemplation. The word **"leisure"** has the connotation of activities, i.e. actions carried out during free time. Leisure activities differ from **tourism** in that they don't require travel for more than a minimum period of 24 hours [...] (RODRIGUES, 2001, p.89).

Through free time, which is part of the capitalist system, tourism capitalizes on this business time by selling spaces created for tourism purposes. These spaces, which are prepared to receive tourists, undergo adjustments to their urban or rural infrastructure, becoming tourist attractions. Spaces equipped with tourist techniques establish new social and spatial relationships, which create new territorialities.

According to Cruz (2001):

Tourism, like other activities - and in competition with them - introduces objects into space that are defined by the possibility of allowing the activity to develop. In addition, pre-existing objects in a given space can also be absorbed by and for tourism, having their meaning altered to meet a new demand for use, the demand for tourist use. (CRUZ, 2001, p.34)

But these relationships and potentialities of tourism are part of the context of globalization and the world in networks and meshes which, through connections and relationships, create and strengthen tourist territories. In relation to territorial formation through meshes and networks, Saquet (2010) states that:

... understands territoriality as multidimensional and inherent to life in society. Man experiences social relations, the construction of territory, interactions and relations of power; different daily activities, which are revealed in the construction of meshes, nodes and networks, constituting the territory; it manifests itself on different spatial and social scales and varies over time. (SAQUET, 2010, p.77)

Through new dynamics that are established in space, new territorialities are acquired, new forms, which unfold over reality, at this point, Saquet (2010) states that:

Territoriality is marked by the movement of appropriation and reproduction of social relations. In this way, the definition of territoriality goes beyond relations of political power, the symbolisms of different social groups and involves, at the same time, economic processes centered on their social agents. (SAQUET, 2010, p.164)

The formation of a tourist territory is related to the insertion of tourist attractions, tourism marketing and the entire infrastructure for receiving tourists. With regard to the conceptualization of tourist territories, Knafou (2001, p. 71) states that "tourist territories - are territories invented and produced for tourism, either through tourism operators or planners". With regard to tourist territories, Cruz (1995) states:

It is a space conquered by tourism, whose peculiar characteristics distinguish it from other territories. One of its main peculiarities, which is also its greatest contradiction, is the absence of defined borders, a basic condition for recognizing the sovereignty of a territory. (CRUZ, 1995, p.5)

These are some conceptions and approaches to the concept of territory, which has become an important analytical tool for geographers, and for our analysis we will use Haesbaert's (2004) conception.

For Haesbaert (2004), the analysis of the territory is established along four territorial lines: the economic territory, which focuses on economic relations in a given space; the political territory, which emphasizes power relations in its spatial dimension; the cultural or Symbolic-Cultural territory, which is formed through the identity and symbolic charge of a territory; and the natural territory, which is constituted through relations between society and the physical-natural environment. With the use of this methodology, the analysis of tourist territories becomes complete and integrated, since tourist territories are endowed with complexity, passing through various aspects, economic, cultural, political and natural.

Transposing the categories of territory proposed by Haesbaert (2004) to the tourist territories proposed by Cruz (1995), since natural, social, economic and political elements are found in the formation of the tourist territory, and delving deeper into the relationships, various relationships can be established with regard to the economic territory and tourism, Tourism is, in principle, an economic activity which is part of the capitalist logic and which, through the relationship between capital and labor, forms tourist territories. Today, it is a gear for the expansion and accumulation of capital, both mobile and fixed.

Tourism, as an economic activity that has been gaining prominence, depends on the formation of a political territory, which is the sponsor of this activity, which through the liberation of public spaces and investments in the sector, makes this activity dynamic and with greater potential to expand capital.

As for the symbolic/cultural territory, this becomes the merchandise of tourism, which consumes the materializations of the culture of a given territory, turning them into tourist symbols through the marketing industry. In this line of thought, on consumption and marketing and the needs of the so-called modern world, Coriolano (2006) states that:

Governments and companies have created an image for tourism and promote it in the *media*. This is necessary because this activity is mediated by the way of thinking, feeling and everything that underpins the culture of bourgeois society. It had to be transformed into a myth, not only by the economic structure of the mode of production, but above all by ideas that transform demand into desire, dreams and aspirations, so an ideology underpins tourism. (CORIOLANO, 2006, p.220)

In this sense, modern society values what is different, which configures various territories full of history and culture into merchandise, strengthening the tourist's imagination, the feeling of pleasure, the experience in these spaces. In this sense, Carlos (2002) mentions:

Leisure in modern society also changes its meaning: from being a spontaneous activity, a search for the original as part of everyday life, it becomes co-opted by the development of a consumer society that turns everything it touches into merchandise, making man a passive element. This means that leisure becomes a new necessity (CARLOS, 2002, p.25).

The last strand focuses on the natural territory, which is characterized by the relationship between society and the physical/natural environment. The characteristics of the natural environment take on new connotations in modern society, transforming its meaning and uses, which imply all the territorial strands mentioned above. Today, the natural environment presents new perspectives and visions regarding its use. Before environmental valuation, forests were seen as capital centered on the timber industry. Today, with latent environmental issues and stricter monitoring, the natural environment has acquired new configurations and uses for capitalism, centered on ecological or adventure tourism, which has the conservation of these territories as its perspective. This new modality has changed social thinking, with the sustainable use of natural resources. Such configurations have been economically valorized, using the various socio-environmental discourses to reinforce ecological and conservation tourism, becoming yet another political and economic cog in society's wheel.

Still in Haesbaert (2004), he emphasizes the hybridity of territories, making their analysis even more complex, by thinking of tourist territories as a great example of hybrid territories, in which there is an exchange between the tourist and the attractions, the local population, where the meeting of the different ends up merging.

There are two conceptions of this analysis, the first of which is positive, as contact enables social exchange and enrichment, but on the other hand tourist territories can lose their characteristics, becoming common spaces.

By using the ideas and methodology proposed by Haesbaert (2004) for the geography of tourism, there is an understanding of the totality of tourist territories and their unfolding.

Panorama, as a territory, has been acquiring formations to become a tourist territory, which presupposes tourist infrastructure supported by tourist attractions and equipment. Panorama can be analyzed from Haesbaert's (2004) theoretical position on territory, with regard to economic aspects, represented by economic changes and the greater representativeness of leisure/"tourism" to the detriment of ceramics and pottery, changes brought about by the implementation of engineering systems, which follow a global logic that reflects on the local. The influence on the territory by political logics, centered on the works carried out by CESP, which serves a development logic of the national territory, but which reflects on the local, altering social and economic relations.

Haesbaert (2004) idealized a different aspect: the symbolic-cultural aspects, which can be transposed to Panorama, because, through the economic changes engendered by the implementation of engineering systems, these changes and impacts caused by the formation of the lake, caused territorial/structural and sentimental/affective losses, which reflect directly or indirectly on society, which lost out on material aspects (infrastructures) and immaterial aspects (feeling of belonging/history). A final aspect of analysis, the natural territory, represented by the formation of the artificial lake, which left part of the municipality of Panorama submerged, reflected in irreversible environmental impacts.

Haesbaert's (2004) theoretical proposal can be framed in Panorama, in view of the changes that have occurred due to global and national logics, which are reflected in the local area. And the new relationships that have been established between people and space, relationships and dynamics imposed by the managers and representatives of CESP, through policies to mitigate environmental impacts, which have (re)ordered the social, economic and environmental arrangements that will be exposed and analyzed in greater detail in the next chapter.

When analyzing tourist territories, the various "views" of tourists and tourism planners must be interpreted. In order to understand the "gaze" on tourism, Urry (2001) emphasizes four central aspects in its constitution: the expectation of a break with everyday life, the way of experiencing the tourist experience (depending on the class and group considered and their respective habits), the search for status and the manipulation of the tourist's gaze. These are the aspects that influence the way tourists look at things, but they are only the main ones. In the light of what Urry (2001) has said, there is an attempt to escape from everyday life, created by the system, and at the same time appropriated by the system, through the commodification and construction of territories. In this line of thought, Urry (2001) mentions:

[...] the tourist gaze, in any historical period, is constructed in relation to its opposite, to non-tourist forms of experience and social consciousness: what makes a given tourist gaze depend on what it contrasts with; what are the forms of non-tourist experience (URRY, 2001, p.16).

And from the point of view and interests of politicians, these complex territories present more than one vision: the vision of the tourist, the vision of the local community and the vision of the businessman, which goes beyond the first two, because their interests are often divergent. For this reflection, Rodrigues (1997) points out:

In Knafou's view, there are two distinct territorialities: that of the local population, identified as "sedentary territoriality", and that of tourists passing through, "nomadic territoriality", indicating competing and contradictory interests that are reflected in the geographical space itself. (RODRIGUES, 1997, p.54)

3. PANORAMA BETWEEN THE "NEW AND THE OLD"

This chapter attempts to present the historical characteristics of Panorama, making a hundred and snow Centrapente, showing the (re)use of urban structures and the natural aspects that shape the present day through historical analysis, which presents the ecological and social relationships that materialize in the space. Panerama has undergone various ecological changes, altering its productive arrangement. In this respect, society is and remains the medley of spaces, and at the same time, it is medleyed by the changes implemented by the ecenemy of the legal and glebal order and by government policies.

Peripheralization allows for a temporal analysis, in which ruggedness is highlighted, in order to understand the dynamics that have taken hold in Panerama and their reflection in the current social, ecological and territorial organization. In this sense, Santes and Silveira (2003) exemplify the importance of peripheralization, stating:

However, a differentiation is necessary because the uses are different in different historical memories. Each periedizaçàe is characterized by different extensions of the form of use, marked by particular interconnected manifestations that evolve together and obey general principles, such as legal history and global history, and especially state and nation (or nations) and, of course, regional features. (SANTOS; SILVEIRA, 2003, p.20)

The territory used acquires new dimensions and new territorialities that refer to historical moments. Periodization opens up the possibility of analysing the use of territory and its manifestations, which govern local, national and even global relations. The relationship between time and space materializes in the forms created by society, merging old and new. The old, which is considered outdated, functions harmoniously with the new. In the case of Panorama, we can see these changes in its urban structure, transport network and economic activities, which have been molded over time to meet the local interests of its representatives/investors. In this vein, Santos and Silveira (2003) mention:

The territory also reveals past and present actions, but already frozen in objects, and present actions constituted in actions. In the first case, places are seen as things, but the combination of present actions and past actions, which the former bring to life,

gives them a pre-existing meaning. This encounter modifies the action and the object on which it is exercised, which is why one cannot be understood without the other. (SANTOS; SILVEIRA, 2003, p.247)

In the territory, the relationships between the fixed and the flows are materialized. In the case of the former, it remains as a structure, but with the changes in the flows, it favours new functionalities, transforming the fixed and becoming part of the present. This movement establishes the notion of pre-existing (old), which signals the changes that have occurred over time, pointing to ways of life and the development of techniques in each period. According to Santos and Silveira (2003):

Solidarities are thus created between new and inherited elements. Ancient forms of storage coexist with modern forms of culture and new forms of transportation and, at each historical moment, mark different technical and social combinations of work. (SANTOS; SILVEIRA, 2003, p.144)

Techniques have changed over time, altering their distribution, speed and access. From a technical point of view, Panorama has changed since its formation, directly influencing the social combinations of work. The formulation of the new in Panorama comes with the essence of the old. The world of work in Panorama has undergone changes, but with the essence and link to the natural environment and the activities that have taken place since its formation, such as fishing, leisure, ceramics, which have incorporated new models.

The gestation of the new in history often takes place almost imperceptibly for contemporaries, since its seeds begin to take hold when the old is still quantitatively dominant. This is precisely why the "quality" of the new can go unnoticed. But history is characterized as an uninterrupted succession of epochs. This idea of movement and change is inherent in the evolution of humanity (SANTOS, 2004, p.141).

Santos (2004) mentions the new na history, relating it to Panorama, which presents the changes that have been driven by economic and political reasons that have generated new territorialities. O new was already present since the idealization of Panorama, because the sustained dynamics and their unfolding in space, and even their materialization, are the result of geographical and natural orders that have been valued over time. The "new" economies that are predominant today already existed, were valued and incorporated new meanings. The "new" meaning of Panorama is guided by tourism, idealized prematurely by CESP and the City Hall. Relationships have changed, the territory is seen in other ways by the city's administrators, as a tourist potential. And there's a different view of the population, which sees the city trivialized by the city hall's management of imaginary tourism.

O tourism currently stands out as one of the main economic activities and is booming worldwide, making this segment one of the main sources of capital accumulation. Despite its magnitude, this service is not accessible to the entire population, but is restricted to the upper classes for luxury tourism and to the new middle class on the rise for mass tourism. Nowadays, with the overvaluation of tourism and the rise in the purchasing power of Brazilians, this activity is becoming more popular, becoming mass, with more accessible destinations that fit in with Brazil's new middle class. According to IBGE data, the travel-related services market generated R$127 billion in 2011. The tourism sector grew from 2.8% of GDP (Gross Domestic Product) in 2009 to 3.6% in 2011.

In Brazil, tourism as an economic activity is relatively recent, and as a characteristic, it appropriates the country's natural areas and historical and cultural riches, transforming them into tourist commodities, as Boullón (2002, p.79) mentions: "Tourist space is a consequence of the presence and territorial distribution of attractions which, we must not forget, are the raw material of tourism". Often, tourist activities are inserted into the space without prior and adequate planning, which ends up harming the population and the environment.

Brazil has great tourist potential, as it has a variety of tourist attractions that are produced using natural and cultural landscapes, which are valued in the modern world. This makes various parts of the country attractive for tourism. With regard to tourism and the use of the landscape, Cruz (2002) states that:

Tourism is the only social practice that fundamentally consumes spaces, and this consumption is through the appropriation of space by tourism, in other words, through the forms of consumption (accommodation services, restaurants, leisure, as well as the consumption of the landscape) that it establishes between tourist and place visited (CRUZ, 2002, p.109).

Given the importance of tourism - which is reflected in the large number of tourists, both Brazilian and foreign, and the financial flows that this sector generates - the segment must be planned and monitored so that it can contribute to the development of the local population, giving them the opportunity to improve their quality of life, achieving social inclusion, while preserving the environment. In this sense, Rodrigues (1997) observes that:

It can be seen that, despite its potential, tourist activity has not been accompanied by adequate planning, nor has it provided space for the participation of the local population, nor has it fostered integration between the various social segments involved. This has created numerous conflicts and obstacles to a sustainable development model (RODRIGUES, 1997, p.56).

Tourism presents itself as one of the main economic activities of contemporary times, idealized and projected in the global and configured in the local, in which the relationships between fixed and flows manifest themselves, crystallizing tourism. Given the data presented, we can see the international influence on this activity, which is valued along the lines of the modern world, where leisure and work culminate in tourism. However, this activity must be planned because it modifies spaces in an ever-increasing attempt to commercialize them. Spaces that have lives, movements, people, feelings, in this sense, we must think about the well-being of local society and its development.

With the rise of tourism, many municipalities have clung to or "appealed" to the idea that their city has great tourist potential. In the case of Panorama it was no different, as they took advantage of the geographical characteristics and its history, which were enhanced by CESP's mitigation works, to label the city as "touristy".

A historical analysis of Panorama is necessary in order to understand the formation of the idealization of a "tourist town" and the products that are valued in this activity, which are the new structures, the roughness and the history of the population. Since this is an activity in which there must be consensus among all those involved, it is about memories and constructions that have been humanized over time. The photographs and history will provide a guide for territorial analysis and the development of Panorama into a leisure area.

CHARACTERIZATION OF THE STUDY AREA

The municipality of Panorama (Figure 1) is located in the far west of the state of São Paulo, on the border with Mato Grosso do Sul, on a landmark centered on the city at 21⁰ 21ꞏ 23" south latitude, and 51⁰ 51ꞏ 35" west longitude. [a]Panorama belongs to the 10th Administrative Region of the state, which has Presidente Prudente as its seat. Currently, the municipality has a population of 14,725 inhabitants according to IBGE estimates (2012).

Panorama borders the state of Mato Grosso do Sul, and this relationship has always been present in its development, with the connection via the Parana River and the ferry, which was used to transport produce and people between the states.

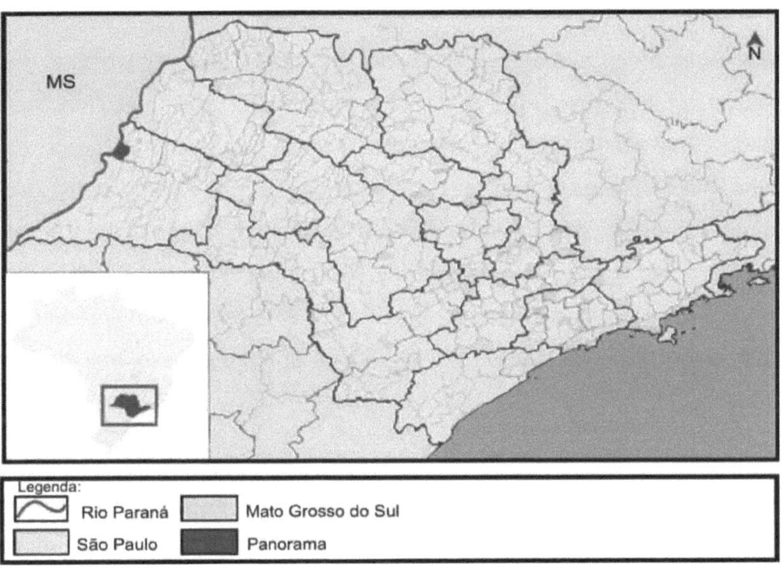

Figure 1- Location of the municipality of Panorama-SP.Org: Tiago

Rodrigues, 2009

According to the Geomorphological Map of the State of São Paulo on a scale of 1:500,000 drawn up by Ross and Moroz (1996), the municipality of Panorama is located in the Parana Sedimentary Basin and on the Western Paulista Plateau. Specifically for the municipality of Panorama, medium and low hills predominate as relief forms, with altitudes ranging from 300 to 480 m, and average slopes between

5% and 20% according to the Geomorphological map of the State of São Paulo.

The Pedological Map of the State of São Paulo (EMBRAPA, 1999) was used to characterize the soils. Focusing on the Panorama region, Red - Yellow Argisols and Red Latosols were identified, the latter being the most representative in the region. These soils are the result of pedogenetic processes that occurred under the Arenitic rocks of the Bauru Group and the basaltic rocks of the Sào Bento Group.

According to the climate classification of the state of São Paulo systematized by Monteiro (1973), Panorama is located in the "west" climate region, with a tropical climate, controlled by equatorial and tropical masses, alternating dry and wet seasons. °Temperatures range from 12 to 35 degrees Celsius, with the annual average being around 26.5 degrees Celsius.

The municipality has a rich hydrographic network, the main course of which is the Parana River and its tributaries - Ribeirão das Marrecas, Itambi, Corrego do Macaco, and of secondary importance is the Peixe River and its tributaries - Corrego da Barranca Funda. With the geographical conditions mentioned, Panorama is in a strategic region, with tourist potential to be exploited, taking advantage of the Parana River for its bathing and fishing, commercializing the spaces recently created in the region that have environmental appeal, such as the Cisalpina Environmental Reserve, created by CESP, located on the south side of Mato Grosso, valuing the existing roughness.

As for the transportation network in which Panorama is located, there is a railroad line that was deactivated in 1998, linking Panorama to the capital of São Paulo. The main highway linking Panorama with other towns is the SP-294 (Comandante João Ribeiro de Barros), which is the most important highway in the region, connecting the entire region. The SP-294 connects the municipality of Presidente Epitacio via Vicinal Lauro Aparecido dos Santos, Ouro Verde via Vicinal PNR-030, Santa Mercedes and Ouro Verde via Vicinal PNR-020 and Paulicéia via Vicinal PNR-10. Air operations are carried out at Dracena airport, which has no flight lines and is considered a small airport.

Figure 2 focuses on the changes that occurred after the formation of the artificial lake on the Parana River, which led to changes in the so-called natural areas and in the urban areas, either because of the flooding or because of the policies carried out by CESP in an attempt to mitigate the impacts on the affected municipalities. The figure drawn up by CESP shows that the reservoir has altered the characteristics of the Parana River, making it wider and affecting urban and rural areas.

Among the municipalities affected in the region shown in Figure 2, Panorama had the most damage in terms of urban area, but in terms of total area, the municipality of Brasilandia-MS lost the most, due to its geomorphology. These changes and dynamics will be discussed in greater depth in the coming chapters.

Figure 2: Panorama region after the lake was formed. Source: CESP

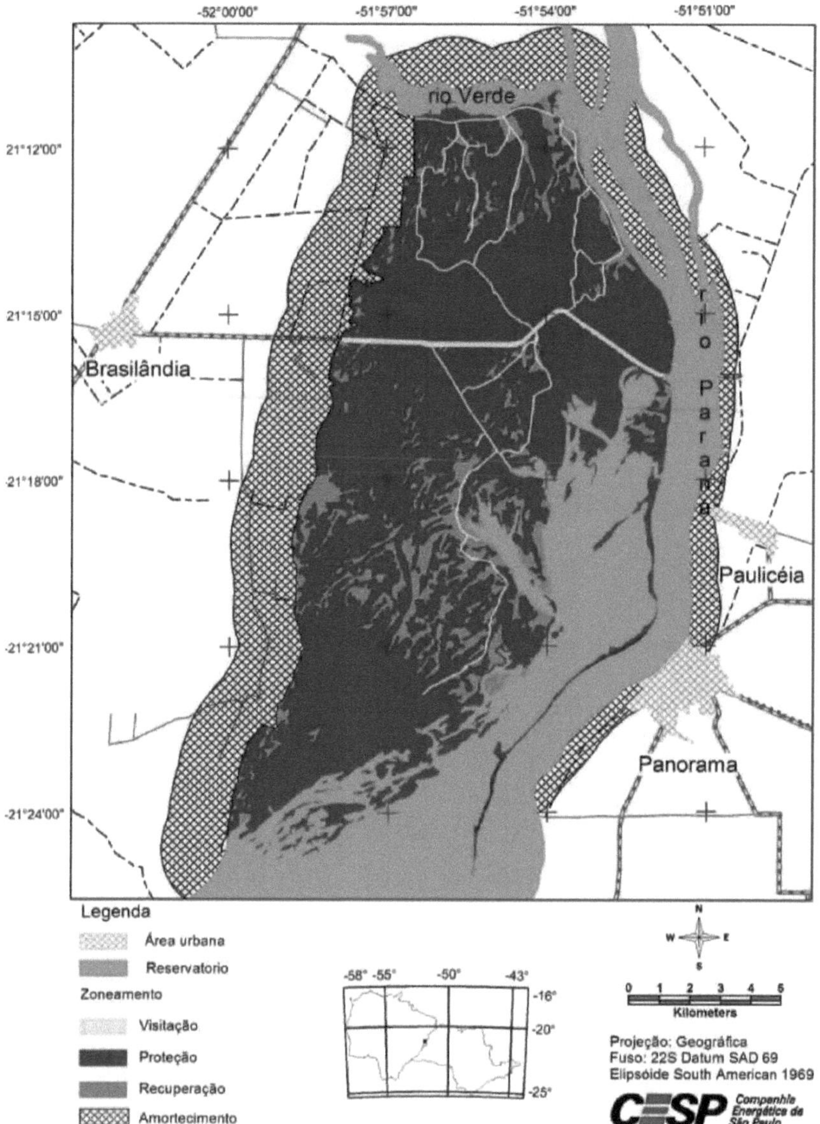

PANORAMA HISTORY

The history and evolution of the municipality of Panorama presents peculiarities that the state of São Paulo itself was going through, and its formation follows in the footsteps of several cities in São Paulo, but presents some particularities. The town of Panorama was created in the municipality of Paulicéia, two kilometers from

Panorama, with land dismembered from the district of Gracianópolis. The development project of the state of São Paulo, centered on the expansion of the Cia. Paulista de Estrada de Ferro railroad network, was the reason for the formation and development of the municipality. Born from the vision and efforts of Brazilian urban planner Prestes Maia, Panorama's history began in 1946, when Quintino de Almeida Maudonnet opened a sawmill (MARQUES 2006).

Through privileged information, Mr. Quintino de Almeida Maudonnet, a traditional businessman and investor from the Campinas family, was informed in 1945 by friends and investors that Cia. Paulista de Estradas de Ferro was planning to extend its railroad network to the border between the state of São Paulo and Mato Grosso, now Mato Grosso do Sul, he decided to form a partnership to buy Fazenda São Marcos Evangelista with 2,700 alqueires next to the Marrecas stream, bordering the Parana River, vacant land owned by Mr. José D'Incao, a pharmacist in Presidente Wenceslau, Alta Sorocabana. Through the purchase of Fazenda São Marcos Evangelista, they set up Imobiliaria Panorama Ltda in December 1945, with Mr. Quintino Almeida Maudon as a partner. Quintino de Almeida Maudonnet, Quintino de Almeida Maudonnet Filho, Arthur Maudonnet, Jùlio Revoredo, Anibal de Andrade, José Ribeiro de Almeida, Guilherme Plichta, Guilherme Rehdder and, as legal consultant, Nelson Noronha Gustavo Filho, Anibal de Andrade, a friend and cabinet officer of the former Mayor of São Paulo, who was invited to visit the region by engineer and urban planner Prestes Maia (MARQUES 2006).

Anibal de Andrade was enchanted by the place, the panorama, the beauty and the potential of the Parana River. He, the urban planner, set out to plan a future city, since, as a Councillor for the Paulista Railway Company, he knew that this would be the end point of the tracks. The plan (Figure 3) of the city was finished and presented in July 1946 by Dr. Prestes Maia, together with an extensive report (Figure 4), at a meeting in Campinas, attended by authorities and people from Campinas society who were pleased with the design of the plan and the oral summary of the report (Marques 2008).

Figure 3: Pianta of the city of Panorama SP 1949: Source:José H. Buzeiin

Figure 4: Report on Panorama-SP: Source: José H. Buzelin

As the town began to form, the company hired loggers to clear the area provided for by the town plan, setting up a sawmill (Figure 5) on the riverbank and the first boards were used to build a hotel, the Rancho Aiegre, the head office and ten small houses for the peasants under the guidance and supervision of the managing partner Guilherme Plichta.

Expenses exceeded forecasts and, in 1948, Mr. Quintino realized that he couldn't continue with the venture (he was already in debt to the loggers and banks). In contact with Dr. Nelson Noronha Gustavo Filho, his legal advisor and President of Companhia Imobiliaria Campineira, the successor to Imobiliaria Campineira Ltda., founded by Mr. Rodion Podolsky, and Dr. Domicio Pacheco e Silva, he suggested that this firm take on the burden and develop the plan. Rodion Podolsky and Augusto Nadalutti, who were responsible for the commercial side of Imobiliaria Campineira, analyzed the problem and came to the conclusion that the only way to solve it would be to transform Panorama Ltda into a public limited company. In order to set up the Sociedade Anónima, it needed subscribers to its shares and, with this, to raise funds.

44

Figure 5: Panorama's first sawmill. Source: José H. Buzelin, 1953.

Mr. Podolsky gathered together eighty-two of the most representative people in Campinas and invited them on an excursion to Panorama. At the Rancho Alegre Hotel (Figure 6), they were introduced to Panorama's potential and the beauty of the area, emphasizing the Parana River, which is approximately 1,200 meters wide (MARQUES 2008).

Faced with this meeting and the presentation of Panorama and its potential as the end city of the railroad network, the majority signed an agreement and accepted the shares of the newly created Panorama S.A., whose Board of Directors, elected and registered with the Board of Trade at the same time, had Augusto Nadalutti as its director.

With changes in the structure of the Panorama S.A. board, Augusto Nadalutti left the management of Imobiliaria Campineira and took on the task of clearing the area and attracting residents to the rural and urban areas. The first step was to pay off the debts owed to the logging contractors (in the annex to the head office). This attitude or the way in which payment was made presented a certain danger, as all the money was placed in a suitcase and the loggers lined up to receive it. The origin of these workers was suspect, as many, living far from urban centers, were fugitives from the police (Marques 2008).

Figure 6: Rancho Alegre Hotel. Source: José H. Buzelin, 1953.

After the formation of the Panorama Real Estate company and the opening up of the area, what had been Patrimonio became a District of the Municipality of Paulicéia, by law no.° .233 of 24/12/1948, a fact that had the help of Federal Deputy José Corrèa Pedroso Junior, who had ties with representatives from Panorama. The cooperation of the Patrimony Administrator, Mr. Antònio Aguiar de Souza, who faithfully carried out the orders received and was also imbued with the pioneering spirit and enthusiasm for the idea of helping to build a city, was of the utmost importance.

As for the isolation, the journey to Tupà, the end point of the Paulista Railroad, 160 km from Panorama to be covered by road, more or less bearable, to Lucélia, then 80 km of roads that were not drivable and sandy areas that required around ten hours by bus to overcome, demanded a lot of effort and unproductive time. The solution found by Augusto Nadalutti to alleviate the distance and the delay was to fly aircraft with a "brevet", which made it possible to build the landing field next to the Ribeirao das Marrecas. The first airplane trips included a stretch between Lucélia and Panorama (Figure 7) over virgin forest with small clearings of the Patrimonios in formation and some gaps denouncing deforestation for coffee plantations. Two years later, the forest had practically disappeared to make way for crops around Adamantina, Flòrida Paulista, Pacaembu, Junqueiropolis, Dracena, Tupi Paulista, all of which were forming and developing, and small towns that had stagnated.

Figura 7: Section between Panorama SP and Lucélia - SP: Source: José H. Buzelin (S/A)

This new form of transportation made it possible to speed up the steps towards the main objective, which was to create the conditions for development that would allow the district to apply to become a municipality in 1953, the year established by law for the territorial subdivision carried out every five years (Marques 2008).

The goal was that in four years the real estate company would create the conditions for the district to achieve administrative autonomy. By December 1949, the estate had changed little. There was a challenge that had to be overcome through the efforts of the administrators and technical and manual labor, which depended on full dedication and detachment. It was a task for many people and, above all, the fulfillment of goals that were centered on populating the countryside and making it productive, which was the first objective and this was achieved with the brokers Yoshimune (Hugo) Matsunaka, Pedro Luiz Nadalutti and Antonio Sapede Filho who scoured Alta Paulista, Noroeste and Sorocabana forming caravans of farmers interested in buying land, mostly settlers who wanted to become landowners, the interested parties were transported by truck and plane to Panorama (Figure 8).

Figura 8: Caravan of land buyers in front of the Panorama hotel. Source: José H. Buzelin, 1953.

It was decided that the remaining area, after the subdivision of the chacaras Circundantes do Patrimònio provided for in Dr. Prestes Maia's project, would be dismembered according to the needs of the traders, with topographical and leasing services carried out by the engineer hired by the settlement company. The measure was validated and seen as important, because it led to the immediate settlement of families and, in less than two years, they were already producing corn, beans and rice planted in the "streets" of the coffee plantations in formation. These measures were of great importance for the formation and elevation of Patrimonio to municipality, given that its location was not privileged, as it was far from the major centers. The leasing of idle land in Patrimonio to plant cotton attracted many interested parties who ended up settling in Patrimonio. This was the step towards the creation of school groups, warehouses, fabric stores, a cinema, pensions, pharmacies, a doctor's surgery, workshops, rice processing machines and, most importantly, the construction of houses.

The brokers Antonio and Luiz Barreto de Oliveira, Adriano Augusto Trondi and Fernando Gardel were in charge of the urban plots. Taking advantage of Dr. Prestes Maia's reputation, they made many sales in São Paulo, Campinas and the interior, which provided the company with an important portfolio of receipts to invest in the development of the Patrimony. The Parana River had many fishermen, all left to their own devices, without any assistance or guidance, and an economy carried out in an amateurish way. The Ministry of the Navy financed the construction, installation and maintenance of the Fishermen's Colony-Ambulatory, whose inaugural ribbon was cut on November 25, 1951, by Congressman Pedroso Jr. who had obtained the funding. That same day saw the inauguration of the electricity generated by a powerful diesel generator, the laying of the foundation stone for the C.P.T. (Cia. Paulista de Transportes) warehouses and C.A.I. offices. C (Companhia Agricola de Imigraçà e Colonizaçà), subsidiaries of Cia. Paulista de Estradas de Ferro, facts that shape the development of Panorama and its power ties, which shaped the will of investors and representatives of that period. (MARQUES 2008)

Panorama began to take on the dimension of a city or the structure of a municipality. The town began to have more buses on the line, people arriving for business or leisure. With the new characteristics that Panorama was acquiring, it was necessary to set up a hotel, which was done through a project by the architects Roberto and Carlos Nadalutti, brothers of Director Augusto Nadalutti, who lived in Rio de Janeiro. The project was offered free of charge and was built under the supervision of Mr. Amador Lombelo, the sawmill concessionaire, the HOTEL PANORAMA (Figure 9), now the Clube Paranoa (Figure 10), with extensions and modified architecture.

Figura 9:   Hotel Panorama (Panorama Company). Source: César Claudino de Souza, 1969.

In the photo above we can see a historical moment in Panorama which signaled a period of development and formation of the municipality. In the photo below we can see Panorama from the current perspective. Instead of the Panorama Hotel, there is now the Hotel Clube Paranoa (Figure 9), which is part of the current tourism practiced, with an emphasis on fishing.

Figura 10: Paranoa Club former Hotel Panorama: Source: Paulo Primo Sobrinho, 2011.

At the time, imports were almost impossible. A tractor was indispensable. For the development and shaping of Panorama's municipal plan, it was always necessary to use economic and political power to fulfill the aspirations of investors and local representatives, and in the context of influencing power, the figure of Dr. Nelson Noronha was decisive. The manager of the Bank of Brazil in Campinas, Mr. Antonio Carlos Bastos, said that he was unable to finance the purchase and authorize the import, considering any attempt by his superiors to be futile. Dr. Nelson and Mr. Augusto went on to Rio de Janeiro and, in the company of Federal Deputies Ferraz Egreja and Pereira Lopes, met with the Minister of Agriculture, João Cleofas, who, in view of the arguments put forward by Panorama's representatives, had a letter drafted for the Bank of Brazil in Campinas, authorizing the financing and import of a D-4 tractor, which landed in Santos in June 1949. With it, the streets were opened up and the airfield was built next to the Ribeirão das Marrecas. This was an important milestone in the development of Panorama, which changed what had been a practically rural landscape (MARQUQES 2008).

[0]On March 28, 1953, the 1st Civil Registry and Notary Office (Figure 11) was set up, headed by Dr.[a] . Aurora Francisco de Camargo. The orientation for the functioning of the registry office was given by the lawyer of Cia. Imobiliaria Campineira, Dr. Enéas

Ferreira Guarita, who was present with his work at the installation of the City Hall, among which he was in charge of the political-administrative organization.

Figura 11:	1st Civil Registry and Notary Office :Source:José H. Buzelin, 1953.

That same year, Panorama's elevation to municipality was requested, a new struggle for this purpose that relied on the power and influence of local politicians and investors, who were almost daily at the Legislative Assembly to talk to deputies asking for support for the initiative. Finally, at the last session of the legislative year, the President of the Assembly, Dr. Rui de Almeida Barbosa, from Campinas, announced the elevation of Panorama to a municipality. With the Municipality approved, it was necessary to register the voters for the election to be held and, to this end, Dr. Nelson de Noronha Gustavo sent his colleague Mr. Antonio Duran to Panorama to register them. In October 1954, the elections were held and Mr. Paulo de Arruda Mendes was elected Mayor, who was sworn in in January 1955. With the emancipation and elevation of Panorama to a municipality, the hopes and intentions of transforming this municipality into the potential of the far west of São Paulo were realized (Marques 2008).

Of great importance was the construction and installation of the railroad network that stretched as far as Panorama, on the banks of the Parana River, which gave the municipality new territories. Called the Paulista West Trunk, a huge branch line from Itirapina to the Parana River was built in 1941 by straightening the lines of three existing branches. From that year onwards, the line, which only reached Tupà, was progressively extended to Panorama, on the banks of the Parana River, arriving in

1962. Passenger trains, operated by Ferroban from November 1998, continued to travel the line precariously until March 15, 2001, when they were discontinued. Panorama station was inaugurated in 1962. The station was provisionally built (Figure 12), out of wood, and ended up staying that way until 1983, when a new station was built, in the same style as the others on the branch, which were from the 1940s. This station was inaugurated in 1984 and today stands as a roughness in the space.

Panorama - July 1982

Figure 12: Old Panorama Railway Station: Source: FEPASA, 1982

Ferroban stopped operating between 2001/2002 and was abandoned along with the entire Bauru - Panorama line, signaling the decline of the rail system in the state along with the closure of stations, including Panorama. The Panorama railroad station was a symbol of the importance of the railroad system and network and of an economic period that was based on the expansion of the coffee industry to the west of the state of São Paulo. At the same time as the railroad guided an economy, it was synonymous with development, and several towns in the west of São Paulo developed close to the railroad network. When the railroad came to a halt, the structures were abandoned, turning into urban roughness that identified a phase of the municipality.

The city as a historically constructed geographical space institutes and guides new forms and functions, as well as solidifying remote forms, whether or not they acquire new functions and forms. Existing roughness inserted into a new socio-spatial context, materializing the past and remaining as a historical mark of a place that once existed, in this sense, Santos (1996) emphasizes:

[11]Let's call roughness what remains of the past as form, built space, landscape, what remains of the process of suppression, accumulation, superimposition, with which things are replaced and accumulated in all places. Roughness presents itself as isolated forms or as arrangements" (SANTOS, 1996, p. 113).

In the context of the railroad station's decommissioning, it was abandoned for five years in decay, becoming a roughness of space, contemplating the past and acquiring a new function in the present. The railroad station is currently used as a municipal library (Figure 13), and the structure has undergone renovations to adapt it to its new functions.

Figure 13: Station turned library: Source: Cristiano Luizăo, 2006.

Panorama has acquired new urban forms, (re)configuring itself and incorporating new territorialities. The old urban spaces that date back to its founding have taken on new forms and uses. Through the roughness that has materialized in space we can see the social and economic changes, understanding the territorialities of today in conjunction with the new economic arrangements.

In the process of urbanization, which materialized in infrastructures, lies the power and influence of local politicians and representatives of that historical moment, which culminated in the formation of Panorama and later its urban fabric. The municipality of Panorama is in the process of restructuring its urban fabric, which brings changes not only to structures, but also to the relationships that are woven into the space, redefining the functions of the city and city life. For Sobarzo (2007):

In this sense, the city is considered to be a product and conditioner of the reproduction of society, of the reproduction of life, of the social relations that are manifested in socio-spatial practice, in other words, the space that is built and modified, in everyday life, in everyday actions, in the use and appropriation that is made of it and, at the same time, the space influencing this everyday life. (SOBARZO, 2007, p.158)

In this sense, it is important to understand the changes and reconfigurations of the spatial arrangement through the historical process because, according to Lefebvre (1968), it is determined by continuities and discontinuities, structures and destructures, evolutions and revolutions over time. Figures 3 and 13 show Panorana in its urban evolution, from the plan that shows the pioneering and clearing of areas to its subsequent position, showing its formed urban network. In this vein of urbanization and its development, Koga (2003) emphasizes the importance of guiding public policies and the relationship between individuals that unfolds over structures.

Thinking about public policy from the territorial point of view also requires an exercise in revisiting the history, the daily life, the cultural universe that lives in this territory, if we consider it beyond the physical space, that is, as a whole range of relationships established between its inhabitants, who in fact build and rebuild it (KOGA: 2003, p. 25).

Figure 14: Image of Panorama-SP. Source: Panorama City Hall, 2010.

After Panorama was elevated to a municipality, various measures were taken to turn the rural town into a proper city. Many structures from the 1950s remain in the municipality, giving it a tourist appeal. Between the old and the new, we can see that several structures have taken on new connotations and uses, changing their meaning and symbolism. These symbols are the roughness of the space that denounce the transition of a socio-economic and spatial model. Among the symbols is the chapel of Sào José, whose project was decreed on July 16, 1958, the construction of the chapel that would be the only large church in the city. On November 15 of the same year, construction began. The chapel, which is owned by the Catholic Church, is open to the public and is one of the town's tourist attractions, according to the town's tourism secretary, Cezar Claudino, in an open interview.

The chapel retains its traditional features from 1959 (Figure 15), but has recently undergone a renovation (Figure 16), despite its excellent state of preservation. Because of the symbolic representation that this church has in the city, it is considered to have strong potential, based on the pilgrimage of the faithful, which is often linked to religious tourism.

Figure 15: São José Chapel. Source: José H. Buzelin, 1959.

Figure 16: São José Chapel. Source: José H. Buzelin, 1959.

Urban life styles are manifested in the city, which reveals the degree to which society interacts with space and its role as a citizen who modifies the landscape. Although Panorama dates back to the 1950s, it already has its roughness, its history, which reveals its urban framework and functionality, supported by the economic sphere that is revealed in the space.

Reflection on the city is fundamentally a reflection on the socio-spatial practice that concerns the way in which life is carried out in the city, as forms and moments of appropriation. Thus, urban space has a profound meaning, as it reveals itself as a condition, means and product of human action (CARLOS, 2004, p.07).

According to Carlos (2004), it is essential to recognize the history of the urban in order to understand the new dynamics and organization of society as a whole, which materializes in space. In an attempt to understand the new dynamics and organization of space and its new meanings, we will contrast the old and the new of parts of the city through photographs.

Figures 17 and 18 show the time of Rua Quintino Maudonett, which today is the commercial hub of the municipality, with a concentration of stores, supermarkets, banks and other commercial figures that are part of the urban area. This is one of Panorama's main streets, since its commerce is limited and concentrated in just a few streets. In the past, this street already had commercial buildings, and its specificity was announced. Space and time are present in the photographs, denouncing the flow that used to be smaller, the urban agglomeration, and the whole question of the technification of urban space.

Figure 17: Quintino Maudonett Street, 1960s: Source: José H. Buzolin, (S/A)

Figure 18: Quintino Maudonett Street. Source: Cezar Claudino de Souza, 2010.

Rodion Podolsky Avenue (Figure 19) can be considered the gateway to the city. Like most of the city, the avenue has changed and intensified its technique in space. This avenue has multiple territorialities and has acquired various uses and specificities.

Figure 19: Rodion Podolsky Avenue :Source: Cezar Claudino de Souza, 2011.

Figure 20: Rodion Podolsky Avenue. Source: Josè H. Buzolin, 1956.

As Figure 20 shows, the avenue was in the process of formation and showed the characteristics of that historical moment of urban formation and development. But even with the technical and temporal changes, several structures remained with the urban rugosities, as is the case with the pharmacy in the photo, the first in the municipality and currently remodeled and renamed Farmacia Dracena. As for the multiple territorialities that this urban space encompasses, it is worth noting that at times when events take place (Carnival / Carnival out of season) it becomes a stage for these events (Figure 21), receiving an average of 20,000 to 30,000 people during each event, according to the tourism secretary, Cezar Claudino, in an open interview. The avenue is home to several bars, snack bars and party services.

Figure 21: Rodion Podolsky Avenue / Carnival: Source: Cezar Claudino, 2008.

The urban will always be on the move and will always crystallize history in its forms, built through the relationship between man and the environment, with the economy as its driving force. But through the frames installed in urban areas, we can observe the past, understand the present and design the future. Carlos (1988) mentions the structure and development of urban movement:

Space is not human because man inhabits it, but because he constructs and reproduces it, turning the object on which work falls into something of his own. On the other hand, space is produced as a function of society's general production process. It is thus a historical product that has undergone and is undergoing a process of technical cultural accumulation, each time displaying the characteristics and determinations of the society that produces it (CARLOS, 1988, p. 15).

Within the urban development of Panorama, we can emphasize the power and interest in real estate speculation, and the desire for the expansion of capital. Panorama can be placed in the context of the settlement of the west of São Paulo, which served the commercial and business interests of that given historical moment. Since the 1950s, with the formal formation of Panorama, this municipality has been acquiring new forms and territorialities, which go beyond the economic.

As far as the transportation system is concerned, Panorama has great potential for navigating the Parana River. On the left bank of the Parana River is the Panorama Intermodal Terminal. It is an Intermodal Terminal, because in addition to the Parana River, the Port of Panorama can be accessed by road (Highway SP-294) and rail (Bauru Regional Unit - UR 3, of Ferrovias Paulistas S.A. - FEPASA). The terminal originated in the 1960s (Figure 22) and has a unique vocation for transhipping solid bulk cargoes such as soybeans, wheat and other products that can be handled by suction and belts (Costa, 1997).

Figure 22: Panorama Intermodal Port in 1968: Source: José H. Buzelin, 1968.

But the original port, which was considered outdated, was decommissioned. Panorama "won" a new intermodal port, built by CESP, as part of the mitigation works for the environmental and social impacts caused by the formation of the Sergio Motta HPP reservoir on the Parana River. The intermodality practiced in Panorama is conditional on the initial transhipment of the waterway barges to the rail cars, but if necessary, it is possible to tranship the cargo to the road trailers.

The Panorama intermodal terminal was very important and active at the end of the 1970s and beginning of the 1980s, with a heavy movement of grain, especially wheat, from Paraná to mills in São Paulo and Jundiai, a fact that coined the name of the terminal.

term "wheat route". With the closure of the railroad network on the stretch from Bauru to Panorama, the terminal fell into decay and has been inoperative since 2006. The most important operations for the port of Panorama were wheat, coming from Guaira - Parana, 453 km to the south, and soybeans, coming from Hernandaryas (Paraguay), Ahrana(2009)

Panorama has undergone major transformations that have marked its history, and through this we can analyze its evolution and development. As we can see from its history, this municipality is similar in its formation to the cities of western São Paulo, but it has always had its own characteristics which are connoted by its geographical aspects. Its location is seen as unprivileged because it is far from the major centers, but it is strategic because it is on the banks of the Parana River, at the end of the important Comandante João Ribeiro de Barros highway, on the border with Mato Grosso do Sul, which is the transport hub of the intermodal system, enabling the flow of production from this region and of people.

The flow between São Paulo and Mato Grosso do Sul was established by ferry, which dates back to the period when the municipality was formed. This was the only means of connection between the states, which hindered the flow, because the ferry had rigid timetables, as well as being expensive to cross, which limited the flow of people from the cities in the region between the states.

The first ferry began operating in Panorama in 1953 (Figure 23). In the historical context, this investment signaled that Panorama could develop and become a hub in the region, because the ferry increased the flow between the states considerably, providing a financial flow. All production in the Brasilandia-MS region passed through the ferry. The state of Mato Grosso do Sul was considered attractive because its land was cheaper than that of the state of São Paulo, causing farmers from São Paulo to invest in this state, mainly in beef cattle and, in this scenario, the ferry brought new perspectives, making it possible to transport the production of Mato Grosso do Sul to the state of São Paulo, which consumed a large part of its production.

In the photographs below, the changes are clear: the first ferry in 1953 shows the flow of people, at this time Brasilandia was less developed than Panorama, the latter being a commercial hub where people sought everyday goods such as fuel, groceries and so on. (Figure 24) from 2000 shows a new phase of the ferry, replaced by the first, which has been losing prestige over time as the municipality of Brasilandia has begun to meet its basic needs, reducing the population flow between the border states. One of the reasons for this decline was the cost of transportation, which meant that it was no longer worthwhile going to Panorama to do your shopping. Other cities larger than Brasilandia have developed nearby, creating new dynamics and ties.

Figure 23: Panorama's first ferry. Source: Cezar Claudino, 1953.

Figure 24: New Balsa: Source: André Kuwada, 2000.

Within the framework of Panorama's urban development, we must analyze the economic aspects since its formation in order to understand the new dynamics that have taken hold in this municipality, altering relations with space and social relations. Panorama's economic development has always been linked to its geographical position, being favored by the Parana River, in this sense almost all economic activities had direct or indirect links with the Parana River. Since the formation of the municipality, various economic activities have influenced Panorama, leaving their roughness in space and altering work and social relations.

As mentioned, Panorama's economic activities have always been linked to the Parana River, denouncing current prospects. Some activities have remained with new arrangements and new territorialities.

Ceramics and pottery were the activities that stood out most in Panorama, reaching their peak in the 1980s with approximately 150 ceramics and potteries in all. This activity was boosted by the large clay deposit in the Parana River. The sector was a highlight in the region, once the first ceramics factory opened in 1953 (Figure 25), a period in which several cities were developing in the region, making Panorama a ceramics hub in the region.

Figure 25: Panorama's first ceramics factory. Source: Cezar Claudino, 1953.

Ceramics were the predominant sector in Panorama until the 1990s, when the ceramics system gradually fell apart, reducing the number of establishments. The decline of several establishments is linked to competition, since ceramics dominated this activity. The potteries have more capital, technology and techniques used in their establishments, and are considered companies, with a large number of employees and a latent production, which makes the market more competitive. Potteries, on the other hand, are artisanal and don't have large investments, making them less competitive on the market, as they are family-run.

In the 1990s several potteries went bankrupt, unable to keep up with the development of ceramics. In 2000, with the formation of the artificial lake, the clay deposits were submerged, leading the sector to decline due to the lack of raw materials. This made the activity unviable in the municipality, as the raw materials were far away, increasing costs. In this new scenario, several ceramics factories closed down, reducing the number of businesses to around fifty, including potteries and ceramics.

Figures 25 and 26 show the old and the new: the characteristics of the ceramics industry have remained almost unchanged, but internally they have been equipped with more modern tools to become more competitive and increase their production. This sector was the one that added the most labor in the municipality and, with the decline of this activity, the workers have joined the new productive arrangements, which will be detailed later.

Figure 26: Ceramics in Panorama: Source: Willian Ribeiro da Silva 2011.

Taking advantage of the Parana River's fishing potential, fishing became one of the main activities for the ceramics industry. In the early days of Panorama, fishing supplied the region and always attracted a large number of fishermen. Given the importance of this sector and the large number of fishermen, in 1952 (Figure 27) the José More de Panorama Fishermen's Colony Z-15 was founded, representing the fishermen's class and serving as a support point. The fishermen's colony acquired new functions and formalities and is now linked to the trade unions to demand better working conditions and guarantees. Currently, the Colony has partnerships with colleges in the region (Faculdades Adamantinense Integrada and Faculdade de Dracena) with courses in zootechnics, biology, among other areas of interest, so that fishermen can get out of amateurism and make this activity more profitable. The members of the Colony are professional and regulated fishermen, with guaranteed rights such as a basic food basket during the piracema period and a minimum wage. Fishermen take part in courses offered by educational institutions and professionals in the field. The Colony has undergone transformations in its structure, as shown in Figures 27 and 28, and is one of the historical symbols of Panorama, representing an activity that has been undergoing changes. With the formation of the lake on the Parana River, it has acquired new dynamics that have changed the role of many fishermen who have become fishing guides, and fishing tourism is considered the municipality's main attraction by the tourism secretariat. The new river dynamics and new working relationships will be developed below.

Figure 27: Fishermen's Colony Z-15 in 1953: Source: Cezar Claudino de Souza

Figure 28: Fishermen's Colony Z-15: Source: Cezar Claudino de Souza, 2010.

Leisure in Panorama, like the other activities mentioned, is linked to the Parana River. This activity has always been present in Panorama, taking advantage of the Parana River's balneability. But leisure in Panorama was not seen by the economy as an activity that did not add value. A specific area was designed by the city council to form the so-called[11] Prainha[11] for residents and visitors.

The area was an attraction for neighboring towns, being well

frequented. The area was nothing more than a forest with an artificial river beach. There was no adequate infrastructure in this space, and visitors brought everything from their homes, from food to barbecues. Figures 31 and 32 show the evolution of the leisure area, which was part of Panorama's history and which has driven new dynamics related to leisure today. To get to the aforementioned "Prainha" it was necessary to cross a bridge (Figures 29) that connected it to this leisure area. With the formation of the lake on the Parana River, it was submerged, and the leisure area was transferred to the Balneario Municipal built by Cesp to mitigate the impacts caused by the formation of the dam.

Figure 29: Access bridge to "Prainha". Source: Cezar Claudino de Souza, 1994.

Figure 30: "Prainha": Source: Cezar Claudino de Souza, 1996.

Figure 31: "Prainha" 1970s: Source: Cezar Claudino de Souza (S/A)

In the historical interweaving of Panorama, we can visualize its dependence on the Parana River and its geographical position. In view of the panorama of activities that have been developed over time, it can be seen that these remain, but with new configurations and functions, highlighting their consequences for society. Urban forms mark periods that tell of the past and guide the present.

The economies that have been established in Panorama are reflected in the urban organization and in the arrangement of relationships that are woven into the urban and rural environment. Currently, the town hall and the tourism office are selling the image of a tourist town and its analysis is essential, based on the historical process of Panorama. In this part of the work, the old was valued, contrasting with the new, in order to understand Panorama's new territorialities and how these weaves unfold in social and spatial relations, characteristics that will be explored in depth in the next chapters, specific visions that will highlight the specificities of Panorama today.

PANORAMA AND ENGINEERING SYSTEMS - The example of HPP. Eng. Sérgio Motta and the Mario Covas Bridge

The changes introduced into Brazilian territory require changes to meet the new economic orders. Given the attempt to modernize the territory, various public policies are designed to modify the territory, making it modern. But public policies are selective and create disparities across the territory, providing some territories with techniques, while others become more rarefied, causing their technical segregation. But this attempt to modernize the territory has the backdrop of making it more fluid for the movement of production on various scales. In this perspective, relating fluidity and transport systems, Santos and Silveira (2003, p.64) emphasize that "there is a demand for rapid movement in the national territory, created by the unification of markets, which is accompanied by greater scope for the actions of firms" in the face of the instrumentalization of the territory by large corporations which absorb the policies implemented in the territory.

In order to make capital more flexible and fluid, it is linked to the materialization of scientific technical objects in the territory, which act in the most diverse territories and localities, giving new meanings and relationships where they are implanted. In this sense, Santos (1996) points out:

The deepening of the division of labor imposes new and more elaborate forms of cooperation and control, on a global scale, where the role of engineering systems designed to ensure greater fluidity of hegemonic factors and greater regulation of production processes, through finance and speculation, is central. (SANTOS, 1996, p.225)

In view of the new economic and social characteristics that Brazil has been experiencing, which are unfolding over the territory through major construction projects, making the territory modern, with new engineering techniques and systems. It is therefore necessary to analyze the role that engineering systems play in shaping the territory and their effects and consequences for the region, focusing on the social and economic aspects, since major works shape the space and its inhabitants.

The concept adopted here is that of "engineering systems", as proposed by Santos (1988, p.79), as being "a set of work instruments added to nature and other work instruments located on top of them, an order created for and by work". Engineering systems focus on the materialization of techniques in space, making it more fluid, through large construction projects, viaducts, highways, hydroelectric dams and the communications sector. Engineering systems are related to the movement of space, through fixes and flows, being the material reaction to the immaterial, which crystallizes in space through major works, which make this space and its surroundings more dynamic.

The fixed, whether natural or social, give rise to engineering systems, which are nothing more than work systems added to nature. In such a way that there are no engineering systems without a transformation of what was previously considered natural. With regard to these systems, it is also worth noting that the more efficient they are, the more they produce in less time, accentuating the division of another factor intrinsic to space, work. Engineering systems modify natural and social structures, shaping and resizing the world of work.

Through these prepositions, Panorama is inserted into the ethics of engineering systems, with major works in this municipality or in the region causing profound changes of various kinds, remembering that national or international dynamics materialize locally through techniques, making territories more susceptible to investments and the fluidity of capital.

The engineering systems set up along the Parana River have changed the dynamics of the region in its various economic, social and environmental aspects. Among the engineering systems set up along the Parana River that have changed Panorama are the Sergio Motta HPP and the Mario Covas Bridge. Sergio Motta and the Mario Covas Bridge, which established new territorial dynamics.

The energy policies of the 1960s and 1970s were based on Brazil's industrial development and the expansion of urbanization. In this context, hydroelectric plants were expanded in the country, with the Parana and Tietê rivers undergoing studies for hydroelectric use, carried out by CESP. In this sense, Gonçalves (1996) emphasizes that:

Since the end of the 1970s, the Paraná basin has assumed importance as a space that is being organized and consumed for the production of energy, as a result of regional and national economic activities determined extremely by the industrial urban process (GONÇALVES, 1996, p.17).

The energy projects carried out from the 1970s onwards are part of the policies to modernize the territory, inserting new productive arrangements into the Brazilian economy. These are national policies that reorganize the regional territory, in the case of the projects on the Parana River, which are an attempt to integrate this area with the rest of the state and insert new development dynamics.

The Sergio Motta HPP was the last and longest project of its kind to be installed in the region, and one of the most controversial of all the power stations built by CESP. It was built on the Parana River, 28 km from its confluence with the Paranapanema River, a region where the hydroelectric potential has been turned to hydroelectric projects. The construction of the Eng. Sérgio Motta HPP, which began in 1979, lasted two decades, and its design underwent changes due to advances in environmental laws in Brazil. Construction began with the support of the São Paulo state government under Paulo Maluf, at a time when in-depth environmental studies were not required. The work was halted several times, characterizing it as a slow and costly project (Figure 32), since several hydro-energy projects carried out by CESP followed the idea of modernization and territorial development, without the environmental appeal, since this energy was considered "clean". However, due to the slowness of the works, the socio-environmental view has changed in Brazil, requiring environmental impact reports, reports and debates with the public.

Society over the implementation of the engineering systems. This project coincided with political and environmental turmoil and several protests were held about its impacts. Data obtained from CESP shows that the Sergio Motta HPP started operating. Sergio Motta HPP began operating in 1999. [2]Its dam, the longest in Brazil, is 10,186.20 m long and its reservoir is 2,250 km long, stretching from the municipality of Rosana -SP to the bases of the Jupia HPP. The reservoir was filled in two stages: the first stage, at 253.00 m, was completed in December 1998 and the second stage, at 257.00 m, in March 2001. In October 2003, generating unit 14 went into operation, bringing the total installed capacity to 1,540 MW.

The data presented by CESP emphasizes the scale of the engineering system implemented in the western region of the state of São Paulo and its repercussions in various cities along the Parana River. (Figure 32) emphasizes the beginning of the works and the detour of the Parana River, showing the power of man over nature and the power play undertaken in the engineering systems aimed at generating energy during that period. This was the last project of its kind and magnitude to be implemented in the region, creating a great deal of controversy in relation to all the projects carried out by CESP, either because of its budget or because of the impact it caused. This work was carried out during a period of environmental emphasis, which generated controversy in the debates between development and the environment.

Figure 32 : Area of the current Sergio Mota HPP. Sergio Mota, start of construction and detour of the Parana River. Source: Tiago Rodrigues, 1982.

Another major engineering project near Panorama-SP is the Mario Covas Bridge (Figure 33), which was completed in 2009 and plays an important role in regional dynamics, as it connects the state of São Paulo, in the municipality of Paulicéia, to Mato Grosso do Sul, in the municipality of Brasilàndia.

Figure 33: Mario Covas Bridge: Source: Willian Ribeiro da Silva, 2011.

This work, although recent, is an old demand of the residents and representatives of the municipalities on the banks of the Parana River and in eastern Mato Grosso do Sul, and of AMNAP (Association of Municipalities of Nova Alta Paulista), since it has been thirty years of struggle for the interstate bridge over the Parana River, which only began its bureaucratic work in 1998 with Governor Mario Covas, whose name was given to the bridge. The Mario Covas bridge is part of CESP's compensation measures for the towns affected by the formation of the reservoir at the Engenheiro Sérgio Motta Hydroelectric Power Station (Porto Primavera). The agreement was signed between CESP and AMNAP on July 3, 1998. Given the complexity of the work, the investment was in the order of 165 million reais, resulting in a bridge that is 1,700m long. The bridge has brought new dynamics to the region, reordering the flow between the two states. A new flow can be seen, especially in the flow of soya and ethanol and in the flow of people, which has become more effective.

THE UNFOLDING OF ENGINEERING SYSTEMS IN PANORAMA

Panorama has undergone various transformations over time, which have given rise to new territorialities at each stage of its history. But the landmark that registers one of its main transformations is the construction of the Sergio Motta HPP and the filling of the artificial lake on the Parana River. Sergio Motta and the filling of the artificial lake on the Parana River, which were the major changes to the landscape and responsible for the formation of a new territory.

With the construction of the Engenheiro Sérgio Motta HPP, Panorama lost 8% of its territory, flooded by the waters of the Parana River, damaging the municipality's main economic activity, ceramic tiles and bricks, because its main raw material, "clay deposits", were under water. Among the changes we can mention the economic sector, since Panorama had the ceramics sector as its main activity, which was "replaced" by tourism. The dynamics that have emerged in Panorama with the formation of the lake are seen by CESP as social and economic development. The ceramics industry has lost its strength due to the lack of raw materials, while CESP has launched urban structures and a bathing resort, which, as a whole, is seen as having great tourist potential by the town hall and CESP. With the formation of the artificial lake on the Parana River, several clay deposits were submerged, affecting the sector, which depends on this raw material, making this activity unfeasible. As a result of the decline of the ceramics sector, CESP, through its mitigation works, created the conditions to establish new activities, which provided new dynamics, such as

emphasized in this research, o tourism. According to the municipality's tourism office, the company has carried out other works that have given it new characteristics at local and regional level, bringing some benefits to the detriment of the environmental impacts caused. As Dias mentions (2003):

On the other hand, the implementation of this type of hydroelectric plant makes it possible to explore other activities, or what is known as multiple use of the potential created, such as navigation, irrigation, tourism, etc. (DIAS, 2003, p.248).

The main product of tourism in Panorama is the Parana River, which attracts tourists from all over the western region of São Paulo and the state of Mato Grosso do Sul, making this activity one of the main drivers of the municipality's economy.

Through the compensatory works carried out by CESP (Chart 3), these have led to the implementation of "tourist" or leisure activities, directly or indirectly, such as the municipal bathing resort, the paving of roads, improvements to the hospital, the construction of a marina, among other works carried out in the countryside, such as bridges, etc.

Table 3: Compensation works in Panorama-SP by CESP

Works:	Panorama - Rio do Peixe road
	Via Marginal
	Renovation and expansion of Santa Casa
	Drainage of two avenues
	Paving and Level Drainage for FEPASA
	Crossing 1 and 2 Ribeirão das Marrecas and Rio do Peixe
	Alternative access Fazenda São José
	Population Relocation
	Sewer Relocation
	Relocation of the Forestry Police Station
	River Port and Railway
	Leisure Area
	Municipal Slaughterhouse (City Hall partnership)

Source: HPP Planning, Engineering and Construction Department.

Sergio Motta: Location of the Works.

The fieldwork showed that the territorial (re)organization in Panorama, as a result of the mitigation works carried out in the area of work, has given fishermen a new role, such as fishing guide (Figure 34), working as day laborers or attached to an inn, which provides services related to fishing and the nautical sector for tourists. This restructuring of work has spilled over into other areas of commerce, through shops specializing in fishing articles, camping, bait, among others, which make up Panorama's "tourism".

Figure 34: Panorama fishing guides. Source: Willian Ribeiro da Silva, 2011.

Over time, the tourist territory in Panorama has acquired new dynamics, which have been established through constructions and the idealization of new spatial forms, which can be represented by the Mario Covas bridge and the residential subdivisions on the banks of the Parana River.

The Mario Covas bridge (Figure 35) is located in the municipality of Paulicéia, approximately 2 km from Panorama, and connects the states of Sao Paulo and Mato Grosso do Sul.

Figure 35: Aerial view of the Mario Covas bridge. Source: César Claudino de Souza, 2008

With the completion of the bridge, a greater flow of traffic was possible in this region, due to the easy access, because before the bridge was built, the crossing between the states was done by ferry. The use of the ferry hindered the interstate flow in this region, because the ferry timetables were inflexible, in addition to the cost of the ferry, thus making it difficult for the population to move across the border of the states mentioned. Through conversations with shopkeepers and the population of Brasilàndia, it can be seen that, with the presence of the bridge, there has been an increase in the flow of people into Panorama, both for leisure and for trade, the latter being in great demand by the population of Brasilàndia for the lower prices and services available that are not found in the municipality of origin.

The bridge has brought new dynamics to the area, as it has increased the flow between the states, opening up the possibility of Panorama receiving a large number of visitors from the south of Mato Grosso, with a wide variety of profiles, as Panorama offers various types of "tourism", with the strengthening of "mass" tourism, or more accessible leisure, with the attractions of using the spa, fishing and nautical tourism. Another important aspect of the bridge is that it is seen by tourism managers and secretaries as a tourist attraction, as it offers a beautiful view of the Parana River, as well as its grandeur, which attracts visitors.

The allotments (Figure 36) followed the dynamics of the bridge. Most of the residential allotments were established near the bridge, thus ensuring a commercial appeal that increased the value of this area and, consequently, the plots in this region.

Figure 36: View of the Aldeia do Lago- Panorama subdivision: Source: Antonio

Rodrigo Mala, 2012.

Allotments in the municipality of Panorama are a new element in the territorial dynamics that date back to the filling of the lake on the Parana River.

With the demand for land on the banks of the river, for the formation of allotments, a speculation was created, which considerably reduced the traditional ranches, those that had the dimensions of farms and ranches and that had direct access to the river, thus reigning the residential on the banks of the river, which is the fractioning of the old ranches, which has complete urban infrastructure, where houses are concentrated that are called "ranches" of medium to high luxury. In this respect, Sanches (1991) mentions:

The increase in demand for leisure and tourist spaces makes investment possible on the part of the real estate sector, which acts to requalify the space and promote its refunctionalization, making it possible to produce residential units in a space in the process of appreciation, where the production of second homes stands out (SANCHES, 1991, P. 35).

As seen above, the allotments are characterized as second homes and are a form of tourism, since a large proportion of the owners do not live in the municipality, but come from different parts of the region, and make use of these spaces for leisure and relaxation, taking advantage of the landscape of the Parana River.

As we have seen, tourism appropriates goods for collective use for its development. In this sense, we can mention the bridge which is becoming a tourist attraction, a public asset which is being used as a symbol of progress, activating real estate speculation in its region on the banks of the Parana River. Both the bridge and the allotments are recent dynamics which are occurring in an intimate way, but which are gradually being incorporated into the space, (re)configuring the territory.

Panorama has undergone various transformations over time, which have given rise to new territorialities at each stage of its history. However, the landmark that registers one of its main transformations is the construction of the Sergio Motta hydroelectric power station and the filling of the artificial lake on the Parana River, which were major modifiers of the landscape and the formation of a new territory. The main product of "tourism", as seen by Panorama's managers, is the Parana River, which, thanks to its bathing water, attracts visitors from all over western São Paulo and eastern Mato Grosso do Sul, making this activity one of the main drivers of the municipality's economy.

The municipal spa symbolizes the effects of the major construction projects on the territory. CESP built the spa, as it did in several towns in São Paulo that were affected by the works, idealizing an imaginary form of tourism, with talk of increasing income and boosting the economy. Panorama is part of this trend, receiving investment and creating false economic images, as is pointed out throughout this article.

Panorama's old leisure area, the currently submerged prainha, had no tourist infrastructure, and only took advantage of the beauty of the Parana River. The new beach, built in a different location from the prainha, has a large number of visitors and is structured with bars, kiosks and aquatic entertainment. A temporal analysis, focusing on the differences and transformations related to the beach, can be seen in Figures 37 and 38.

Figure 37. Former bathing resort of Panorama-SP. Source: César Claudino de Souza, 1990.

Figure 38: Current Panorama-SP bathing resort. Source: César Claudino de Souza. 2006.

In the figures above, it's clear to see the change in the structures of the Panorama municipal spa, which is now one of the main "tourist" attractions, where concerts and events are currently held in the spa grounds, which have infrastructures that can accommodate large events.

The traditional events (discussed in more detail below) held in Panorama moved from the clubs to the beach, gaining new dimensions and becoming regional events, promoting the town and changing the view of Panorama in the region, which had previously only been known for the ceramics trade. These are some of the changes following the implementation of the Sergio Motta HPP. The construction of the power station and the formation of the lake were the main agents of these changes, which took place around mitigation works and agents from the private sector.

LEISURE AND TOURISM IN PANORAMA

Panorama has an intrinsic relationship with the environment because it has been part of the Parana River landscape since the town was founded, and this environment has therefore boosted the local economy. Among the economic activities in the municipality that have been marked by the presence of the river, we can mention the establishment of potteries, livestock and fishing. Recently, Panorama has been excelling in the practice of tourism and its main attraction is the landscape of the

Parana River. Leisure used to be present in Panorama, through the little beach which attracted visitors from all over the region. With the construction of the Balneario, and the technification of the space, leisure has acquired a new dimension, seen from an economic point of view. Other tourist attractions in the town are sport fishing and events that are held annually and festivities on commemorative dates.

In order for Panorama to become attractive for the practice of "tourism" or the expansion of leisure, investments were made by the private and public sectors, mainly in infrastructure and equipment related to tourism, giving new characteristics to the municipality's economy and (re)configuring the territory, in an attempt to crystallize a new economic activity idealized by the municipality's managers.

Tourism currently stands out for its socio-economic potential and its ability to produce and transform spaces, turning them into merchandise, making it the new gear of capital accumulation. Tourism is characterized by its ability to absorb public and private investment, which produces new configurations affecting space and society. From this perspective, it can be said that tourism and leisure have space as their main commodity, transforming it so that it becomes attractive, according to Cruz (2000):

No other activity consumes space as elementally as tourism does and this is an important factor in differentiating tourism from other productive activities. It is through the process of consumption of space by tourism that tourist territories are created (CRUZ, 2000, p.17).

Currently, due to the prominence of tourist activity in Panorama, the municipality has been undergoing various transformations that affect its productive process, resulting in a production and (re)production of the territory.

Panorama's economy was for a long time linked to livestock farming and the production of bricks and tiles. In the 1980s, ceramics activities expanded considerably in the municipality, due to the large amount of clay on the banks of the Parana River; this was the activity that most employed the local population. Nowadays, the ceramics sector is losing its economic importance, while tourism stands out as a possibility for the town's economic and social development.

In order to develop tourist activity, a municipality needs attractions such as infrastructure and tourist facilities, as defined by Boullon (2002, p.54) "facilities include all establishments run by the public authorities or by the private sector and dedicated to providing basic services". From this point of view, tourist activity must be planned and have policies that control and direct this activity, providing local development and the participation of the whole of society in this process, so that it does not prioritize individual interests. In this vein, Cruz (2002) mentions:

The way in which tourism appropriates a certain part of the geographical space depends on the public tourism policy that is implemented in the area. Public tourism policy is responsible for establishing goals and guidelines to guide the socio-spatial development of the activity, both in the public sphere and in private initiative. In the absence of public policy, tourism takes place in absentia, i.e. at the whim of private initiatives and interests (CRUZ, 2002, p.09).

With the construction of the Sérgio Motta hydroelectric power station in the municipality of Porto Primavera, which transformed the Parana River into a large artificial lake, there was consequently interference in public and private property, the road system, electricity distribution, sanitation, leisure and, above all, the environment.

Faced with the set of negative impacts caused in the regional context by the implementation of hydro-energy projects, CESP and every company, as an entrepreneur, is obliged to minimize and mitigate these effects as much as possible for the environment and society (DIAS, 2003, p.242).

Among the mitigating works, CESP built the "Frederico Platzeck" Municipal Beach, which has become one of Panorama's tourist attractions. It has an artificial beach, measuring 18,000.00 m^2 with capacity for 1,300 bathers. According to the municipality's tourism office, the company has carried out other works that have given it new characteristics at local and regional level, bringing some benefits to the detriment of the environmental impacts caused. As DIAS (2003) mentions:

On the other hand, the implementation of this type of hydroelectric plant makes it possible to explore other activities, or what is known as multiple use of the potential created, such as navigation, irrigation, tourism, etc. (DIAS, 2003, p.248).

In the territorial modification of Panorama, in the constitution between the "old and the new", visions of the implantation have also changed, because the population has lost part of its history, humanized structures that materialized in space, its relationships with space and, many, its source of income. Through open-ended interviews with the population, the feeling of belonging to the affected areas can be observed, as well as to the course in which Panorama is inserted. The works and compensation have not recovered the lost history and activities that were part of Panorama. On in-ioco visits, it can be seen that the works imposed by CESP do not boost the economy, and are seen by residents and traders only as palliative measures, or a simple "stop-gap" for the damage that has been done. CESP creates a vision of development and modernization with the construction of some improvements, which in its ideology can generate employment and the emergence of new economic dynamics that lead to social development. On its website, CESP emphasizes the "works in the community" from the point of view that they have led to development, but leaves in the background the emotional and material losses that were laden with a sense of belonging. CESP's perspective is that:

Throughout its history, CESP has developed a program to mitigate socio-economic impacts in the areas of influence of its hydroelectric plants, benefiting farmers, ranchers and fishermen in the communities. Many localities have been created or modernized as a result of the company's actions. (Site: www.cesp.com.br/portalCesp/Visitado on 28/12/2012 at 11:00)

Among the changes that have produced new characteristics for the territory and for tourism in Panorama is the construction of a bridge (Figure 35), located in Paulicéia, 2 km from Panorama, which connects the states of Mato Grosso do Sul and São Paulo, increasing the flow in this region, incorporating new

The bridge is also an attraction for tourism.

In view of the transformations that have taken place in Panorama, especially in the socio-economic and environmental sectors, which were motivated by the construction of the Engenheiro Sergio Motta Hydroelectric Power Station and the formation of the artificial lake, the importance of scientific and academic study in this area is shown, in order to understand the production of the territory and its reality in its various spheres, for better planning and territorial management.

Among the investments made by the public sector are infrastructure works in an attempt to establish tourism in the municipality. These works were carried out by the São Paulo Energy Company (CESP) as compensatory and mitigating measures for the impacts caused by the construction of the Sergio Motta hydroelectric power station and the formation of the artificial lake. Through this item, it can be seen that the incentives provided by public authorities and articulated organizations are related to "tourist" activities in the municipality of Panorama.

PANORAMA'S MAIN TOURIST ATTRACTIONS

In order to develop tourism activities, the municipality must have tourist attractions that enable the flow and permanence of tourists and visitors who will consume the spaces and attractions, boosting the economy, with tourist attractions being the driving force behind this activity. According to the Ministry of Tourism (2007), tourist attractions include:

Tourist attractions - places, objects, equipment, people, phenomena, events or manifestations capable of motivating people to visit them. Tourist attractions can be natural; cultural; economic activities; programmed events and technical, scientific and artistic events. (Ministry of Tourism, 2007 p. 28)

Tourism in Panorama has as its main attraction the Parana River (Figure 40), which is considered a tourist commodity, as defined by Souza (2003). This transformation of nature into a commodity by tourism, with the prospect of greater profit, has been carried out with voracity, compromising the main commodity itself, nature, and in the future, the unfeasibility of the tourist activity itself.

Figure 39- Aerial view of the municipality of Panorama-SP. Source: César Claudino de Souza, 2005.

These aspects show the importance of the Parana River in the practice of tourism and the possibility of bringing local development to the population. This activity must be planned so that all segments participate, so that it is developed responsibly and does not compromise the environment, among others.

Sport fishing is another activity that is developed in Panorama, becoming an attraction that is intrinsically linked to the river. With the formation of the artificial lake, the natural conditions of the river change, making sport fishing more popular. Several species of fish have been introduced along the Parana River, which are traditional for this activity, such as *Cichla monoculus*, the scientific name for the popular tucunaré, an exotic species from the Amazon basin. To meet this demand, there are a large number of specialized guides who know the river well and the strategic points for fishing, as many of them were former fishermen. As well as guides, the town has several stores specializing in fishing articles, a marina and inns.

Panorama's artisans have been gaining space and expression with the formation of the artisans' cooperative, giving local artisans the opportunity to show their art, transposing the culture and characteristics of the municipality and the region in which it is located. In an open interview with the president of the Casa do Artesão, Marival Ferreira Costa, she mentioned that the cooperative was founded in 2002, and that until 2010 the spaces were rented, relying on the collaboration of the artisans and the town hall to keep the building rented. In November 2010, the Casa do Artesão's own headquarters was inaugurated (Figure 40), a project carried out with funds from the federal government, built by the municipal administration, with two bathrooms, one for the disabled, a kitchen, a room for courses and exhibiting canvases and a larger room for exhibiting and selling the various handicrafts.

Figure 40: Casa do Artesão de Panorama: Source: Marivai Ferreira Costa, 2011.

The Casa do Artesão currently has 20 members, making various handicrafts made by the members themselves, including clay crafts, river stones, roof tiles, wood, biscuit, webs, crochet, embroidery, among others.

The strength of the artisans and their organization in cooperatives, together with the small traders, is reflected in the attempt to valorize local activities and their representations, making it possible to enter the local and regional market, which often becomes specialized. In this vein, Coriolano (2006) mentions:

Small places and (or) enterprises find a way to enter the tourism production chain with local products or goods. The contradictions in tourism are more evident than in other economic activities because it is, in its origins, elitist, produces so-called non-places, sometimes denies the local and degrades cultures in order to maximize profit. (CORIOLANO, 2006, p. 45)

One of the main tourist attractions is the municipal spa, built by CESP as a compensatory measure. The "Frederico Platzeck" Municipal Spa has an artificial beach (18,000.00 m^2) with a capacity for 1.333 bathers per period), a shower complex, public lighting, main gate, toilets, 3 snack bars, 6 large and 2 small kiosks, 1 restaurant, an events square with a stage for concerts and capacity for 15.000 people, fishing pier, mooring, boat and jet-ski loading and unloading ramp, boat shed, camping area with support facilities, 1 hostel with 10 apartments, sports square with:

3 multi-sports courts, 2 sand soccer pitches, 2 beach volleyball courts, 1 *society soccer* pitch, 1 bocce court, parking for cars and buses, 2.300 m long, *playground* and administration building, data provided by the tourism office.

The resort is the scene of events that take place in the municipality, such as the annual carnival, which is organized by the town council. Previously, this carnival was held in clubs, but with the decline of these clubs, it is now held in the resort, attracting visitors from all over the region, with around 60,000 people attending over the course of the five-day festival.

There is also Pan Verão, which is an off-season carnival organized by private initiative. Pan Verão, which takes place in November and lasts three to four days (weekends), is based on the original carnival. This event also attracts large crowds and is held at the municipal bathing pool.

The crossing of the Parana River, which has become a traditional event started by Mr. José Gonçalves (1966) (Figure 41) through information that there was a young man who swam across the river with

great facilities attracting many swimmers to the competition.

Figura 41: 1ª Crossing the Parana River from Panorama. Source: José H. Buzelin, 1956.

Today, the crossing (Figure 42) is considered to be the second largest event held in the municipality of Panorama. This event began with one start and 25 swimmers; today it is organized with three starts and has already had 300 swimmers taking part, according to the Secretary of Tourism. The event has acquired more technical equipment, such as a referee and monitoring with nautical equipment, ensuring greater safety for the competitors. The event used to take place in the old "Prainha", but now it's held in the municipal baths. The crossing can be contrasted historically through the figures, which emphasize its potential to attract visitors from the region, making it one of the tourist attractions listed by the tourism office.

Figura 42: Crossing the Parana River from Panorama in 2011: Source: Cezar Claudino de Souza

The feast of the town's patron saint, St. Joseph, attracts many devotees to receive his blessings, which are distributed over the course of nine days (called the novena). During the festival there are charity cattle auctions, masses and processions organized by the Parish Pastoral Council and the Parish Economic Affairs Council, thus involving the whole town.

Among the festivities considered attractive by the tourism office is the Popcorn Festival celebrated on St. John's Day.

It is an event offered by the Town Hall, held in the Praça do Povo, with the aim of bringing together all the local population, with free distribution of popcorn, peanuts and quentào, as well as performances by children's and adult gangs.

The concerts held in the Praça do Povo, which has an exclusive stage for the performances, events funded by the town hall, are considered by the local administration to be a way of bringing the population together and developing a meeting point, becoming another attraction for the region. The New Year's Eve festivities are also events under the responsibility of the town hall, with live concerts in the town's spa and fireworks at New Year's Eve. These events have become major attractions, receiving tourists from all over the region and from other states, showing the importance and potential of the municipality for "tourism" and leisure.

The tourist attractions presented were mentioned by the municipality's tourism management. It can be seen that several of the attractions and events mentioned do not lead to tourism as such, but rather to leisure. Panorama can be considered a potential that needs to be exploited and which requires investment and a tourist dimensioning if it is to become a reality.

According to the figures presented by SEBRAE-SP, Panorama has six inns geared towards fishing tourism, offering everything from lodging to food and the entire fishing sector. Some inns are geared towards fishing and leisure, thus forming a structure to cater for tourists who are looking for leisure, tranquillity and a family atmosphere. In an interview with Maria Suzumura, owner of Pousada Aquàrio, she emphasized the need for investment in leisure and tourism in Panorama, in order to attract a greater demand from tourists. According to the businesswoman, several inns are investing in structures to attract the fishing family (tourists) as a whole. Another fact that permeates tourism and leisure in Panorama is its seasonality, depending on the period of the piracema, in this sense the inns are investing in swimming pools and aquatic attractions to ease the discrepancy of tourists during the year. This view became general since other owners interviewed had the same concerns.

According to the SEBRAE survey, Panorama also has five hotels and a campsite, as well as twenty-one eateries, including restaurants, snack bars, bakeries, pizzerias and beer bars.

Work in Panorama has also become seasonal, since demand depends on the fishing season. The jobs that are considered permanent are generally not linked to the tourist workforce. Those who work directly with tourism depend on the flow of tourists, which is seasonal. Panorama has 69 fishing guides and 20 artisans who make their living from tourism, always at risk due to the instability of the activity developed in Panorama. With the closure of several potteries and ceramics plants, the only alternative for the population was to boost their practices and add them to "tourism". In this sense, we can see that the idealized tourism in Panorama is on the fringes of the marketing promoted by CESP and its managers, and the lack of an option to invest in another sector, or even to attract investment.

Panorama is characterized by the elements presented above as being in a consolidation phase, since tourism in Panorama arose spontaneously, without proper planning, since the process of touristification happened gradually. Most of the town's tourist facilities were acquired after the Sergio Motta hydroelectric power station was built. Leisure activities in Panorama have been carried out since its formation, always linked to natural assets such as the river, forests and others, which have been boosted by managers as an economic alternative for the municipality.

The spontaneous process of tourism formation is based on activities that already existed, but which have been given a new guise by marketing. Through open-ended interviews with the tourism secretary, it can be seen that there is a great effort to turn the activity into a potential, or it can even be seen as an exaltation of the activities and nature that already belonged to Panorama. The brochure (Figure 43) produced by the municipality's tourism office shows that the activities being promoted are prerequisites that have always been present in Panorama, such as fishing, bathing in the river and taking advantage of the natural beauty. In this sense, it can be said that "tourism" in Panorama was born spontaneously, due to its natural characteristics.

Tourism marketing takes advantage of the consumer world to sell places and sensations that are contemplated today. In this sense, there is an overvaluation of the natural, rustic environment, which escapes the urban geometries of the big cities, and Panorama's administrators have added values and marketed nature and its attributes. The town hall created the slogan "The most beautiful sunset" (Figure

44) at the entrance to the city, which reinforces the role of tourism marketing and the vapourization of nature, turning it into a commodity.

Figure 44: Entrance to Panorama and its Tourist Slogan: Source: Wiiiian Ribeiroda Siiva, 2012.

In view of the new tourist perspectives and the creation of tourist spaces and the consumption of existing and natural ones, Panorama fits into this prism of creating tourist environments enhanced by the power of marketing. Corioiano (2006) highlights the power of consumption and the influence of tourism *marketing*:

From a market perspective, tourism *marketing* is versatile, as it induces consumption, makes places into marketable products, creates expectations of easy wealth for entrepreneurs, with the argument of offering a variety of work options (CORIOLANO, 2006, p. 58).

Figure 43: Folder promoting tourism in Panorama; Source: Secretarla turismo, 2010.

In this respect, there are various elements that need to be developed in Panorama in order to achieve real tourism. For example, the preparation of activities to welcome tourists according to the type of tourism the municipality is looking for. Cobra mentions this:

Understanding the demand for tourism services presupposes a range of types, from the origin of the tourism, the reason for the trip, the means of transport used, the geographical characteristics of the destination, the life cycle of the destination (is it growing, maturing or declining), the duration of the vacation, the profile of the tourist group, the type of accommodation sought, the pattern of spending, to who organized the trip. (COBRA, 2002, p.69)

As Cobra (2002) emphasizes, leaders must pay attention to the smallest details in order to achieve professional tourism, which will bring sustainable and lasting development to the municipality. In this respect, Panorama is beginning to make progress, as it already has a board of directors dedicated to the municipality's tourism activity, a cooperative and an association focused on matters relating to the activity, which are gradually creating the basis for developing "professional" tourism.

In terms of sustainable development, with an emphasis on the socio-economic and environmental, Panorama has experienced various environmental impacts resulting from the formation of the artificial lake on the Parana River, which has given new configurations to the economy and social relations, which has unfolded over the territory, which in a contradictory way has created a new "possibility" and "tourist" structure in the municipality. Through the impacts came various infrastructure works, as a compensatory measure, which gave new dynamics to the space, as well as forming a humanized landscape that shaped the Parana River in a new way, and the possibility of new species of fish, which are suitable for sport fishing, which is one of the attractions of tourism in Panorama. According to Veloso (2003):

The starting point is the environment, where strong legislation must be established and enforced, which is fundamental for the development and maintenance of tourism activities. At no time, whatever type, modality, segmentation or form of tourism is developed, can the preservation and conservation of the environment not be the main part of tourism development. (VELOSO, 2003, p.84)

The environmental aspect is of great tourist appeal to Panorama, as several attractions are related to the natural landscape, so it is important to conserve it and raise awareness among all segments of society. With environmental concerns, environmental education should be intensified, emphasizing tourism, since for many municipalities this is the only source of income, so preservation is necessary. In this sense, there is training in all segments, with the concern to train conscious professionals.

Training, as mentioned above, can take place through official programs, such as the National Program for the Municipalization of Tourism (PNMT), developed by Embratur, as well as through professional or technical teaching entities, such as universities, colleges, SEBRAE, among other publicly recognized entities (VELOSO,

2003, p.89).

Panorama is a long way from the professionalization of leisure or "tourism", due to the lack of a tourism perspective and investment in courses to train the people who depend on this sector. Tourism in Panorama has amateurish characteristics, in terms of receptivity, service and development.

Activities linked to leisure and tourism in Panorama absorbed a large part of the workers who were employed directly or indirectly in ceramics or pottery. However, this transition was compulsory, due to the lack of options. The workers felt this transition without undergoing adequate training, which led to the precariousness of the services sold in Panorama.

Considering all of the above, it shows the importance of public agents in overseeing tourism activity so that it is carried out in a sustainable and planned way, in order to achieve development without damaging the environment. The role of planning and monitoring must be followed by society as a whole, with an awareness of the need to preserve the environment, which is the driving force behind the activity that guarantees the survival and maintenance of the municipality.

In the light of the fieldwork carried out during the high season (open fishing) and the low season (piracema) with open interviews with the population, there were reports that there is a lack of investment in the sector and preparation of all those involved in this activity. Among the complaints is the town hall's emphasis on carnival events, which tarnishes the town's image, being known mainly for these events that don't add value and don't boost the town's economy, according to residents. The residents interviewed stressed the importance of valuing festivities and traditional events that attract visitors with their families, such as the crossing of the River Paranà, because, in their view, these events can boost the town's economy.

In the view of the traders, there is a lack of unity between the city council and the trade to guide investments in the sector and to plan and annual calendar of activities that attract tourists and visitors throughout the year, pointing out that Panorama's tourist activities are seasonal. As for the events, there were complaints about the carnivals, because at the end of the carnivals several structures are damaged, and it is an activity that does not boost trade, since the goods consumed come from the tourist's place of origin or even from visitors.

There was a consensus on the view of CESP and its works, as both residents and traders reported that the only positive aspect was the construction of the bathing resort. They also mentioned the losses related to the compensation paid by CESP, and that the works did not repair the material and emotional losses.

CESP's vision emphasizes development and the generation of new economic dynamics that enable the inclusion of the population and its socio-economic growth. The vision of the ideal implemented by CESP becomes contradictory to the realistic vision of society. In view of the works, we can emphasize that CESP obeys the national political orders that are part of the attempt to make the Brazilian territory more dynamic in order to attract capital.

In this paradigm, major engineering works are the consequences of a global system that reflects on the local, and the local population does not benefit from the positive aspects of these works, leaving them to deal with the ills of this system and the negative impacts. In this scenario, Panorama finds itself inserted, impacted and visualized by CESP as a city that has gained from its works. But in reality it has lost its historical characteristics, its economic activity and its relationship with the natural and economic environment.

In order to comfort or alleviate the damage, "tourism" was introduced, which CESP saw as a great potential and that the spa alone would be an attraction that would boost the city's entire economy, and this ideal was taken up by the city's administrators and managers. This reality was not only introduced in Panorama, but in several cities impacted by CESP.

In conversations with residents, they don't see tourism in Panorama, which can be listed as the same activities, such as bathing and fishing, the lack of an adequate urban structure with all the tourist facilities that a so-called tourist town should have, and the professional mentality to add value and attract tourists.

In the traders' view, there is a lack of commitment on the part of the town hall to really make tourism viable, with the implementation of equipment and its maintenance, making the town an attraction. For them, the main attraction that could make the town a tourist destination is sport fishing, which should be promoted and publicized in an attempt to make Panorama a tourist destination, boosting sport fishing.

In an open interview with the municipality's tourism secretary, Cezar Claudino, he mentioned Panorama's potential, and that according to the managers, Panorama qualifies as a tourist city. He also mentioned that Panorama is competing in the state of São Paulo to win the title of tourist agency, and that these bureaucratic procedures are in their final stages, which will boost Panorama's tourism.

In view of the visions presented, it can be seen that Panorama is faced with various paradigms and a contradictory reality. The new dynamics inserted into the territory come through external economic manifestations, of which the people of Panorama do not take advantage. Tourism or the expansion of leisure activities came from CESP's idea, as the municipality had no choice but to reorganize itself economically and include its entire population. Even though Panorama has great potential for tourism, it is marginalized, without adequate infrastructure and excluding part of the population.

4. CONCLUSION: TOURISM IN PANORAMA?

In the contemporary world, leisure and tourism are becoming increasingly important on the economic and social scene. They can be seen as a way of bringing development to the community involved, but at the same time they can have an impact and increase social inequalities. In this sense, it is necessary for all elements of society to participate in the structuring and planning of tourist activity, since this, seen from a social perspective and articulated with all segments, can raise the quality of life of the population and territorial organization. The relationships that are established in space obey a global logic that crystallizes in the local, since this dynamic is not available to everyone, and in this sense, this movement can create spaces of exclusion. This movement can include tourism, considered one of the main economic activities today, capable of expanding capital through the commodification of spaces.

In the face of the worldwide appreciation of tourism, Panorama, through its new territorial dynamics, including CESP's mitigation works and their social, environmental and economic consequences, has created the idea that it is a tourist town. This vision, boosted by the managers of the territory, together with CESP and its works, is an attempt to commercialize spaces and activities that were already present in Panorama. As analyzed in this paper, Panorama was impacted by the formation of the artificial lake, which led to the decline of its main activity, ceramics and pottery. From this perspective, with CESP's vision of development at the expense of the impacts caused, the managers are trying to turn Panorama into a tourist town, in order to absorb the workforce that used to work in ceramics and pottery before the impacts.

As proposed in this work, fieldwork was carried out at different times, given the seasonal nature of Panorama's activities, with open-ended interviews for greater detail and knowledge of the area, and for data collection. The fieldwork was the basis of the research, allowing us to go beyond the readings and demystify the reality of tourism in Panorama, and how it is seen by different segments of society. The demystification comprises the visions and discourses about the activities developed in Panorama, verifying whether tourism is just an ideal constructed by managers. The first fieldwork involved interviews with the leaders of the municipality's tourism department, which made it possible to understand the strategies, and the collection of photos and documents that helped the research. During the first contact, the secretary spoke about the events and the difficulties encountered in the work, emphasizing the need for investment in infrastructure and equipment related to tourism. There is a lack of "tourist" awareness among shopkeepers regarding procedures with tourists, customer service, information and services. These impressions were verified through open-ended interviews with traders. In order to better develop the activities linked to tourism, as mentioned in this paper, the sector needs professional training courses and professionals assigned to planning in various areas in order to get to know the study area, emphasizing the attractions and the structure in general.

The second piece of fieldwork focused on open interviews, formal and informal conversations with local traders and tourists. The businesses chosen are directly or indirectly related to tourism, and range from fishing tackle to snack bars and restaurants. The shopkeepers reported that the flow of visitors to Panorama influences the movement of the establishments, emphasizing fishing, which is considered the main attraction of the municipality, activating the local economy. Traders reported that the flow of visitors increased with the completion of the Mario Covas bridge, as this provided greater interaction between the states. With regard to the town hall, the majority of traders mentioned the lack of incentives and planning related to tourism. In their view, the city government should hold training courses, invest in and revitalize the structures that are considered tourist attractions, and expand the promotion and marketing of Panorama. However, the municipality's interventions are centered on events, as the interviewees reported. With regard to events, the traders complained about the carnivals, which are not economically attractive for the municipality because the public brings food and everything they need to spend the period of the event from their place of origin. Another fact mentioned by traders and the local population is the loss of family character in carnival events, attracting a negative image of the municipality in the eyes of tourists and towns in the region, which could affect other segments of "tourism" and leisure in Panorama, such as fishing and family leisure, which is in search of natural landscapes and tranquillity. As for the improvements that need to be made in Panorama, according to the traders and the population, there were several opinions, with the emphasis on the construction of a tourist portal, the construction of a tourist information center, maintenance of the spa, children's playgrounds, a gym for the elderly, investments in attractions, and in the aesthetics of the municipality, cleaning and maintenance of municipal structures, these being some of the demands. The population and tourists/visitors pointed out that the opening hours of the shops are inflexible, being closed at times when the town receives a larger flow of tourists/visitors.

As for the open-ended interviews with "tourists" / visitors, there was a diversification of the profile of tourists in Panorama. Regarding the city of origin, most of the tourists/visitors interviewed came from the west of São Paulo and the east of Mato Grosso do Sul, accompanied by their families or on excursions, characterizing them as day tourists, according to Andrade (2004). The majority of tourists/visitors got their information about Panorama from friends. These tourists/visitors usually eat in restaurants, stay in inns, or return the same day to their city of origin. Among the suggestions for improving the municipal park, the tourists mentioned the installation of toilets in the kiosks, live music and more attractions. As already mentioned, the landscape of the Parana River is the town's main attraction, and this was the main reason that attracted most of the interviewees.

It should be noted that Panorama has undergone major transformations as a result of the implementation of engineering systems, the construction of the Sergio Motta HPP and the Mario Covas Bridge, which have provided new elements that have reordered the territory, emphasizing "tourism" and leisure. The construction of the Sergio Motta Hydroelectric Power Station and the Mario Covas Bridge provided new elements that reordered the territory, emphasizing "tourism" and leisure. This made it possible to restructure leisure in the municipality with mitigating works, which ensured new meanings and visions for commerce, work and the local economy in general.

Based on the interviews and fieldwork, together with the theoretical foundation, it can be seen that Panorama lies between tourism and leisure. The tourism practiced in

Panorama is centered on fishing, as tourists stay overnight in inns and consume, which is classified as tourism by the UNWTO (2001). On the other hand, the greater flow of visitors is reflected in the use of the municipal spa as a leisure activity. In this activity, visitors don't consume in the municipality, they stay for just one day, and in this perspective the WTO doesn't consider it to be a tourist activity, in the words of Andrade (2004) they are visitors, and the practice is leisure. Along these lines, Rodrigues (1997) points out:

In terms of leisure travel, the most significant phenomenon in the state of São Paulo, from the point of view of the number of people who travel, are the Sunday excursions, the demand for which is represented by members of the economically less privileged social strata of the population, pejoratively known as "farofeiros". (RODRIGUES, 1997, p.119)

Panorama is undergoing a process of "touristification", according to Dantas (2007), which focuses on the use of space over a certain period of time that has consequences in the socio-cultural sphere, manifesting itself in the place, emphasizing the transformations made by human action for leisure purposes.

Panorama has great potential for structuring "real" tourism activity. This is underway with some measures and manifestations, such as the creation of the tourism secretariat, and the inclusion of Panorama in the Circuito Turistico Oeste Rios, organized by SEBRAE, in an attempt to publicize the tourist potential of western São Paulo.

As an expression of this activity, we can see the association of artisans and fishermen, which has a direct link to tourism. The fishermen have been taking courses to meet the demand for fishing, becoming fishing guides. These initiatives could give Panorama the label of tourist town, with the practice of professional tourism, which could lead to local development in a sustainable way.

The work presented shows that tourism must be thought of from a social point of view, so that it leads to the development of all the people involved in this activity, but in order to do this, it is necessary for the managers of the territory to engage in planning and policies that make this activity viable in a conscious way.

5. BIBLIOGRAPHY

AHRANA - **Parana River Waterway Administration.** *Data and Information Report, 2008.* São Paulo, 2009.

ALENTEJANO, P. R. R. and ROCHA-LEÂO, O. M. **Fieldwork: an essential tool for geographers or a trivialized instrument.** Boletim Paulista de Geografia, São Paulo, n⁰ 84, p. 51-57. 2006

ANDRADE, José Vicente. **Tourism, fundamentals and dimensions.** 7.ed. São Paulo: Ed. Atica, 2004.

BENI, Mario Carlos. **Structural analysis of tourism.** São Paulo: SENAC, 2008

BERTONCELLO, Rodolfo V. **Tourism and large metropolises: the city of Buenos Aires.** In: RODRIGUES, Adyr Balastreri (Org.). *Turismo e geografia* : reflexôes teóricas e enfoques regionais. São Paulo: Hucitec, 1996. p. 209-223.

BULLON, Roberto C. **Planning the tourist area.** Bauru: Edusc, 2002.

CARACRISTI, Isorlanda. **Tourism as it is done and the development we want.** In: CORIOLANO, Luzia Neide (Org.). Turismo com ètica. Fortaleza: UECE, 1998.

CARLOS, Ana Fani Alessandri. **Urban space: new writings on the city.** Sao Paulo: Contexto. 2004

. YAZIGI, Eduardo e CRUZ, Rita de Cassia Ariza da. (Org). **Tourism - space, landscape and culture.** 3. ed. São Paulo: Hucitec, 2002.

. **Tourism and the production of the non-place.** In: YAZIGI, Eduardo; CARLOS, Ana Fani Alessandri; CRUZ, Rita de Cassia Ariza da. (Orgs.). *Tourism*: space, landscape and culture. São Paulo: Hucitec, 1996.

. Ana Fani A. **The (re)production of urban space.** São Paulo: Editora da Universidade de São Paulo, 1994.

COBRA, Marcos.**Marketing de turismo.**São Paulo:Cobra,2002.

CORIOLANO, Luzia Neide. **Inclusive tourism.** Fortaleza: FUNECE, 2007.

. Tourism and sustainable social development. Fortaleza: EDUECE, 2003.

CRUZ, Rita de Cassia Ariza da. **Public tourism policies in Brazil:** meaning, importance, interfaces with other sectoral policies. In: Maria José de Souza. (Public policies and the place of tourism. Brasilia: UnB/MMA, 2003.

. **Tourist hospitality and the urban phenomenon in Brazil**: general considerations. In: Célia Maria de Moraes Dias. (Hospitality: reflections and perspectives. 1 ed. São Paulo: Manole, 2002.

. **Introduction to the geography of tourism**. São Paulo: Roca, 2001.

Tourism policy and territory. São Paulo: Contexto, 2000.

. **Tourism policies and the construction of the tourist-coastal space in the Northeast of Brazil.** In: Amalia Inés Geraiges de Lemos. (Org.) Tourism: socio-environmental impacts. São Paulo: Hucitec, 1995.

COSTA, L.S; SILVEIRA. *Inland waterways in Brazil.* Brazilian Navy Documentation Service, 1997.

DIAS, Jailton. **The construction of the landscape on the São Paulo - Paraná - Mato Grosso do Sul border:** a study by remote sensing. 2003. Thesis (Doctorate in Geography). Unesp, Presidente Prudente.

EMBRATUR. **Tourism concepts.** EMBRATUR: Department of Economic Studies, Division of Tourism Economics, Brasilia, 1992.

HAESBAERT, Rogério. O mito da (des)territorialização: do fim dos territórios à multiterritorialidade. Rio de Janeiro: Bertrand Brasil, 2004.

IBGE. Database: Available at: < http://www.ibge.gov.br.> Accessed in2012.

OLIVEIRA, W. **The socio-environmental impacts of the Porto Primavera HPP in the municipality of Anauralàndia - MS**. 2004. 171 f. Thesis (Doctorate in Geography) - Universidade Estadual Paulista/FCT. Presidente Prudente - SP.

UNWTO. **Sustainable tourism development: manual for local organizers**. Brasilia: UNWTO, 1994.

LOURENÇO, C. **Fieldwork**: the geographer's laboratory par excellence. Revista Geografia Passo a Passo - Ensaios Criticos dos Anos de 1990. Presidente Prudente, 1991.

MARQUES, J. B. **Heroic Times**. Dracena: Editora Jòia, 2008.

. Pioneers: **People of fiber**. Dracena: Editora Jòia, 2006.

MINISTRY OF TOURISM. **Roteiros do Brasil. Tourism Regionalization Program. Operational Module 7: Tourist Routes.** Brasilia: 2007.

. **National Tourism Plan 2007/2010**: a journey of inclusion. Brasilia: MTur, 2006.

. **National Tourism Plan**: guidelines, goals and programs 2003/2007. Brasilia, 2003

MOESCH, Marutschka Martini. **The production of tourist knowledge**. São Paulo: Contexto, 2000.

MONTEIRO, C. A. de F. A. **A dinàmica climatica e as chuvas no Estado de Sao** *Paulo.* São Paulo, USP Geography Institute, 1973.

MORAES, Antonio Carlos R. **Meio Ambiente e Ciéncias Humanas**. São Paulo: Annablume, 2005.

PETROCCHI, Mario. **Tourism**: planning and management. 2. ed. São Paulo: Futura. 1998.

GONÇALVES, Carlos W. P. "**Geografia Politica e Desenvolvimento Sustentavel**", in RevistaTerra Livre. 1996.

RAFFESTIN, Claude. **Towards a geography of power**. Translated by Maria Cecilia. França. São Paulo: Atica, 1993.

Rizzo, Marçal Rogerio. **Encounters and mismatches between tourism and sustainability:** A study of the municipality of Bonito - Mato Grosso do Sul. 2010. Thesis (Doctorate in Geography). UNESP, Presidente Prudente.

RODRIGUES, Adyr Balastreri. **Geography of tourism: new challenges**. In: TRIGO, Luiz Gonzaga Godoi (Org.). **Turismo**. How to learn, how to teach. 2. ed. São Paulo: SENAC, 2001

. **Tourism and local development**. São Paulo: Hucitec, 1997.

. **Tourism and the environment, reflections and proposals**. São Paulo:

Hucitec, 1997.

. **Tourism, modernity and globalization**. 1st edition. São Paulo: Hucitec. 1997.

. **Tourism and Geography - Theoretical Reflections and Regional Approaches**.

São Paulo: Hucitec, 1996.

ROSS, J. L. S & MOROZ, I. C. **Geomorphological map of the State of São Paulo**. Journal of the

Department of Geography, São Paulo, n.10, p.41-56, 1996.

SAQUET, M. A. **Approaches and conceptions of territory**. São Paulo: Expressão Popular, 2010.

SANCHES, J.E. **Espacio, economia y sociedad. Madrid**: Sigio

Véintiuno, 1991.

SOBARZO, Oscar. **Notes on a theoretical-methodological proposal for analyzing public spaces in medium-sized cities.** In: Sposito, M.E.B (Org). Medium-sized cities: Spaces in transition. Sào Pauio, Expressào Popular, 2007.

SANTOS, Milton. **Divided Space:** The Two Circuits of the Urban Economy of the Underdeveloped Countries. São Paulo: EDUSP, 2004

. Milton and SILVEIRA, Maria Laura. **Brazil**: territory and society at the beginning of the 21st century. Rio de Janeiro: Record. 2003

For another globalization: from single thought to universal consciousness. São Paulo: Record, 2000.

. **Space and method**. 4. ed. Sào Paulo: Nobel, 1997.

The nature of space: technique and time, reason and emotion. São Paulo: Hucitec, 1996.

The return of the territory. In. SANTOS, M., SOUZA, M. A. A. and SILVEIRA, M. L. (eds), Territory: globalization and fragmentation. Sào Paulo: Hucitec/Anpur, 1994.

For a political economy of the city. São Paulo, Hucitec, 1994.

Metamorphoses of inhabited space. Sào Paulo : Hucitec, 1988.

. **Space and Method**. Sào Paulo: Nobel, 1985

. **Space and society**. Petropolis: Vozes, 1982.

SOUZA, Marcelo Lopes and RODRIGUES, Glauco Bruce. **Urban planning and**

social activism. São Paulo: UNESP, 2004.

SOUZA, E. B. C. Nature and consumption - the contradictory relationship of sustainability in tourism. **Revista Oiència Geografica** (AGB/Bauru). Bauru. Ano IXvol. IX, n.3, p. 253-258, Sep/Oct 2003.

SUERTEGARAY, D. M. A. **Field research in Geography**.

GEOgraphy (UFF), Niterói/RJ

TAVARES, Adriana de Menezes. **Oity Tour.** São Paulo: Aleph, 2002.

TRIGO, Luiz Gonzaga. **How to learn tourism, how to teach** it. São Paulo: Senac, 2001.

URRY, John. **The tourist's gaze**: leisure and travel in contemporary societies. São Paulo: Studio Nobel/SESC, 2001.

VELOSO. Marcelo. **TOURISM, simple and efficient** Ed.Roca: 2003, São Paulo-SP

YAZIGI, Eduardo (org.) **Tourism and landscape.** São Paulo: Contexto, 2002.

More
Books!

info@omniscriptum.com
www.omniscriptum.com
OMNIScriptum

Printed by Books on Demand GmbH, Norderstedt / Germany